Parts of Classes

Parts of Classes

David Lewis

With an appendix by

John P. Burgess, A. P. Hazen,
and David Lewis

Basil Blackwell

Copyright © David Lewis 1991

First published 1991

Basil Blackwell Ltd
108 Cowley Road, Oxford, OX4 1JF, UK

Basil Blackwell, Inc.
3 Cambridge Center
Cambridge, Massachusetts 02142, USA

All rights reserved. Except for the quotation of short passages for the purposes of criticism and review, no part of this publication may be reproduced, stored in a retrieval system, or transmitted, in any form or by any means, electronic, mechanical, photocopying, recording or otherwise, without the prior permission of the publisher.

Except in the United States of America, this book is sold subject to the condition that it shall not, by way of trade or otherwise, be lent, re-sold, hired out, or otherwise circulated without the publisher's prior consent in any form of binding or cover other than that in which it is published and without a similar condition including this condition being imposed on the subsequent purchaser.

British Library Cataloguing in Publication Data

A CIP catalogue record for this book is available from the British Library.

Library of Congress Cataloging in Publication Data

Lewis, David K.
Parts of classes/David Lewis.
p. cm.
Includes bibliographical references.
ISBN 0-631-17655-1 — ISBN 0-631-17656-X (pbk.)
1. Set theory. 2. Whole and parts (Philosophy) I. Title.
QA248.L48 1991
511.3'22 — dc20   89-49771
                                                    CIP

Typeset in 12 on 14pt Bembo
by Graphicraft Typesetters Ltd., Hong Kong
Printed in Great Britain by Billings & Sons Ltd., Worcester

Contents

Preface   vii

1 Taking Classes Apart   1
   1.1 Fusions and Classes   1
   1.2 Classes and their Subclasses   3
   1.3 Are there any other Parts of Classes?   6
   1.4 The Null Set   10
   1.5 Consequences of the Main Thesis   15
   1.6 More Redefinitions   17
   1.7 Sethood and Proper Classes   18
   1.8 A Map of Reality   19
   1.9 Nominalistic Set Theory Revisited   21

2 The Trouble with Classes   29
   2.1 Mysterious Singletons   29
   2.2 Van Inwagen's *Tu Quoque*   35
   2.3 On Relations that Generate   38
   2.4 Quine on Urelements and Singletons   41
   2.5 The Lasso Hypothesis   42

Contents

2.6	Ramsifying out the Singleton Function	45
2.7	Metaphysics to the Rescue?	54
2.8	Credo	55

3 A Framework for Set Theory — 61

3.1	Desiderata for a Framework	61
3.2	Plural Quantification	62
3.3	Choice	71
3.4	Mereology	72
3.5	Strife over Mereology	75
3.6	Composition as Identity	81
3.7	Distinctions of Size	88

4 Set Theory for Mereologists — 93

4.1	The Size of Reality	93
4.2	Mereologized Arithmetic	95
4.3	Four Theses Regained	98
4.4	Set Theory Regained	100
4.5	Ordinary Arithmetic and Mereologized Arithmetic	107
4.6	What's in a Name?	109
4.7	Intermediate Systems	112
4.8	Completeness of Mereologized Arithmetic	113

Appendix on Pairing — 121
  by John P. Burgess, A. P. Hazen, and David Lewis

Index — 151

Preface

There is more mereology in set theory than we usually think. The parts of a class are exactly the subclasses (except that, for this purpose, the null set should not count as a class). The notion of a singleton, or unit set, can serve as the distinctive primitive of set theory. The rest is mereology: a class is the fusion of its singleton subclasses, something is a member of a class iff its singleton is part of that class. If we axiomatize set theory with singleton as primitive (added to an ontologically innocent framework of plural quantification and mereology), our axioms for 'singleton' closely resemble the Peano axioms for 'successor'. From these axioms, we can regain standard iterative set theory.

Alas, the notion of a singleton was never properly explained: talk of collecting many into one does not apply to one-membered sets, and in fact introduces us only to the mereology in set theory. I wonder how it is possible for us to understand the primitive notion of singleton, if indeed we really do.

In March 1989, after this book was mostly written, I learned belatedly that its main thesis had been anticipated in the

Preface

'ensemble theory' of Harry C. Bunt.[1] He says, as I do, that the theory of part and whole applies to classes; that subclasses are the parts of classes, and hence that singletons – unit classes – are the minimal parts of classes; and that, given the theory of part and whole, the member-singleton relation may replace membership generally as the primitive notion of set theory. There is some significant difference from the start, since Bunt accepts a null individual and rejects individual atoms. But the more important differences come later: in my discussion of the philosophical consequences of the main thesis, and in our quite different ways of formulating set theory within a mereological framework. I think there is difference enough to justify retelling the story.

Later in 1989 I had another surprise. I learned how we can, in effect, quantify over relations without benefit of the resources of set theory. (All we need is the framework: plural quantification and mereology.) John P. Burgess discovered one way to do it; A. P. Hazen discovered a different way. It was not quite too late to add an appendix, jointly written with Burgess and Hazen, in which these methods are presented and applied. They open the way to a new 'structuralist' interpretation of set theory, on which the primitive notion of singleton vanishes and my worries about whether it is well understood vanish too.

Good news; but not a full answer to my worries. The set theory of the future may go structuralist, if it likes. But I can't very well say that set theory was implicitly structuralist all along, even before the discoveries were made that opened

---

[1] Harry C. Bunt, *Mass Terms and Model-Theoretic Semantics* (Cambridge University Press, 1985), pp. 53–72, 233–301.

Preface

the way. The set theory of the present – which means the bulk of present-day mathematics – rests on the primitive notion of singleton. Is it rotten in its foundation? I dare not make that accusation lightly! Philosophers who repudiate all that they cannot understand have very often gone astray. Maybe my worries are misguided; maybe somehow, I know not how, we have understood the member-singleton relation (and with it membership generally) well enough all along. If so, structuralism is a real solution to an unreal problem, and we might as well go on as before.

The book reflects my indecision. I whinge at length about the primitive notion of singleton (section 2.1), but then I mock those philosophers who refuse to take mathematics as they find it (section 2.8). The book mostly proceeds under the working assumption that singleton is a legitimate primitive notion. But also it presents the structuralist alternative, in section 2.6 and in the appendix. (What's said taking singleton as primitive will not in any case go to waste. All but section 4.6 admits of structuralist reinterpretation.) Structuralism is at least a very welcome fallback. But I would much prefer a good answer to my worries about primitive singleton, so that I could in good conscience take mathematics as I find it.

I thank all those who have helped by discussion of this material, especially D. M. Armstrong, Donald Baxter, Paul Benacerraf, George Boolos, Phillip Bricker, John P. Burgess, M. J. Cresswell, Jennifer Davoren, Hartry Field, A. P. Hazen, Daniel Isaacson, Mark Johnston, Graham Oppy, Hilary Putnam, W. V. Quine, Denis Robinson, Barry Taylor, and Peter van Inwagen. I thank Nancy Etchemendy for preparing the figures in the appendix. I am indebted to the Melbourne Semantics Group, where this material was first presented in 1984; to Stanford University, where

Preface

a more developed version was presented as the 1988 Kant Lectures; to Harvard University, for research support under a Santayana Fellowship in 1988; and to the Boyce Gibson Memorial Library.

David Lewis
Princeton, January 1990

1

Taking Classes Apart

1.1 Fusions and Classes

Mereology is the theory of the relation of part to whole, and kindred notions.[1] One of these kindred notions is that of a mereological *fusion*, or *sum*: the whole composed of some given parts.

The fusion of all cats is that large, scattered chunk of cat-stuff which is composed of all the cats there are, and nothing else. It has all cats as parts. There are other things that have all cats as parts. But the cat-fusion is the least such thing: it is included as a part in any other one.

It does have other parts too: all cat-parts are parts of it, for instance cat-whiskers, cat-cells, cat-quarks. For parthood is transitive; whatever is part of a cat is thereby part of a part of the cat-fusion, and so must itself be part of the cat-fusion.

The cat-fusion has still other parts. We count it as a part of itself: an *improper* part, a part identical to the whole. But also

---

[1] See sections 3.4 and 3.5 for further discussion of mereology.

Taking Classes Apart

it has plenty of *proper parts* — parts not identical to the whole — besides the cats and cat-parts already mentioned. Lesser fusions of cats, for instance the fusion of my two cats Magpie and Possum, are proper parts of the grand fusion of all cats. Fusions of cat-parts are parts of it too, for instance the fusion of Possum's paws plus Magpie's whiskers, or the fusion of all cat-tails wherever they be. Fusions of several cats plus several cat-parts are parts of it. And yet the cat-fusion is made of nothing but cats, in this sense: it has no part that is entirely distinct from each and every cat. Rather, every part of it overlaps some cat.

We would equivalently define the cat-fusion as the thing that overlaps all and only those things that overlap some cat. Since all and only overlappers of cats are overlappers of cat-parts, the fusion of all cats is the same as the fusion of all cat-parts. It is also the fusion of all cat-molecules, the fusion of all cat-particles, and the fusion of all things that are either cat-front-halves or cat-back-halves. And since all and only overlappers of cats are overlappers of cat-fusions, the fusion of all cats is the same as the fusion of all cat-fusions.

The *class* of all cats is something else. It has all and only cats as members. It has no other members. Cat-parts such as whiskers or cells or quarks are *parts of* members of it, but they are not themselves members of it, because they are not whole cats. Cat-parts are indeed members of the class of all cat-parts, but that's a different class. Fusions of several cats are *fusions of* members of the class of all cats, but again they are not themselves members of it. They are members of the class of cat-fusions, but again that's a different class.

The class of As and the class of Bs can never be identical unless the As are all and only the Bs; whereas the fusion of the As and the fusion of the Bs can be identical even when none of the As is a B. Therefore we learn not to identify the class of As with the fusion of As, and the class of Bs with the fusion of Bs, lest we identify two different classes with one single fusion.

A member of a member of something is not, in general, a member of it; whereas a part of a part of something is always a part of it. Therefore we learn not to identify membership with the relation of part to whole.

So far, so good. But I used to think, and so perhaps did you, that we learned more. We learned to distinguish two entirely different ways to make one thing out of many: the way that made one fusion out of many parts, versus the way that made one class out of many members. We learned that fusions and classes were two quite different kinds of things, so that no class was ever a fusion. We learned that 'the part-whole relation applies to individuals, not sets.'[2] We even learned to call mereology 'The Calculus of Individuals'!

All that was a big mistake. Just because a class isn't the mereological fusion of its members, we shouldn't jump to the conclusion that it isn't a fusion. Just because one class isn't composed mereologically out of its many members, we shouldn't jump to the conclusion that there must be some unmereological way to make one out of many. Just because a class doesn't have all and only its members as parts, we shouldn't jump to the conclusion that a class has no parts.

1.2 Classes and their Subclasses

Mereology does apply to classes. For classes do have parts: their subclasses. Maybe they have other parts as well; that remains to be seen. But for now, we have this

---

[2] So said David Lewis in *Philosophical Papers*, vol. I (Oxford University Press, 1983), p. 40. I might at least have granted that the part-whole relation applies to classes in a *trivial* way: even if a class has no proper parts, as I then thought, at least it should have itself as an improper part.

First Thesis: One class is a part of another iff the first is a subclass of the second.

To explain what the First Thesis means, I must hasten to tell you that my usage is a little bit idiosyncratic. By 'classes' I mean things that have members. By 'individuals' I mean things that are members, but do not themselves have members. Therefore there is no such class as the null class. I don't mind calling some memberless thing – some individual – the null *set*. But that doesn't make it a memberless class. Rather, that makes it a 'set' that is not a class. Standardly, all sets are classes and none are individuals. I am sorry to stray, but I must if I am to mark the line that matters: the line between the membered and the memberless. (Or I could impose novel coinages on you, which would probably be still more annoying.) Besides, we had more than enough words. I can hijack 'class' and 'individual', and still leave other words unmolested to keep their standard meanings. As follows: a *proper class* is a class that is not a member of anything; a *set* is either the null set or else a class that is not a proper class; and an *urelement* is any individual other than the null set.[3]

My First Thesis, therefore, has nothing to say yet about the null set. It does not say whether the null set is part of any classes, nor whether any classes are part of the null set. I shall take up those questions later. Now that you understand what the First Thesis means, what can I say in its favour?

First, that it conforms to common speech. It does seem natural to say that a subclass is part of a class: the class of women is part of the class of human beings, the class of even

---

[3] Warning: I introduced a *different* idiosyncratic usage of 'class' in my book *On the Plurality of Worlds* (Blackwell, 1986), pp. 50–1n.

Classes and their Subclasses

numbers is part of the class of natural numbers, and so on.[4] Likewise it seem natural to say that a hyperbola has two separate parts — and not to take that back when we go on to say that the hyperbola is a class of x-y pairs. To a German, the First Thesis might seem almost tautological: a standard word for 'subset' is '*Teilmenge*', literally 'part-set'. Sometimes, for instance in the writings of Cantor and Zermelo, the word is just '*Teil*', sometimes with an implication that the subset is proper, or proper and non-empty. (Here I am indebted to Ignacio Angelelli.) The devious explanation of what we say is that we speak metaphorically, guided by an analogy of formal character between the part-whole relation and the subclass relation. The straightforward explanation is that subclasses just are parts of classes, we know it, we speak accordingly.

Second, the First Thesis faces no formal obstacles. We learned, rightly, that membership could not be (a special case of) the part-whole relation because of a difference in formal character. But the subclass relation and the part-whole relation behave alike. Just as a part of a part is itself a part, so a subclass of a subclass is itself a subclass; whereas a member of a member is not in general a member. Just as a whole divides exhaustively into parts in many different ways, so a class divides exhaustively into subclasses in many different ways; whereas a class divides exhaustively into members in only one way. We have at the very least an analogy of formal character, wherefore we are free to claim that there is more than a mere analogy.

Finally, I hope to show you that the First Thesis will prove fruitful. Set theory is peculiar. It all seems so innocent at first! We need only accept that when there are many things, then

---

[4] As noted in D. M. Armstrong, *Universals and Scientific Realism* (Cambridge University Press, 1978), vol. II, pp. 36–7.

also there is one thing – the class – which is just the many taken together. It's hard to object to that. But it turns out later that this many-into-one can't always work, on pain of contradiction – yet it's just as hard to object to it when it doesn't work as when it does. What's more, the innocent business of making many into one somehow transforms itself into a remarkable making of one into many. Given just one single individual, Magpie or Possum or the null set or what you will, suddenly we find ourselves committed to a vast hierarchy of classes built up from it. Not so innocent after all! It is exactly this ontological extravagance that gives set theory its welcome mathematical power. But, like it or not, it's far from what we bargained for when we first agreed that many can be taken together as one. We could understand set theory much better if we could separate the innocent Dr Jekyll from the extravagant and powerful Mr Hyde. The First Thesis will be our first, and principal, step towards that separation.

1.3 Are there any other Parts of Classes?

The First Thesis leaves it open that classes might have other parts as well, besides their subclasses. Maybe classes sometimes, or always, have individuals as additional parts: the null set, cat Magpie, Possum's tail (and with it all the tail-segments, cells, quarks, and what not that are parts of Possum's tail). To settle the question, I advance this

Second Thesis: No class has any part that is not a class.

The conjunction of the First and Second Thesis is our

Are there any other Parts of Classes?

Main Thesis: The parts of a class are all and only its subclasses.

But the Second Thesis seems to me far less evident than the First; it needs an argument. And an argument needs premises. My premises will be the First Thesis, plus three more.

Division Thesis: Reality divides exhaustively into individuals and classes.

Priority Thesis: No class is part of any individual.

Fusion Thesis: Any fusion of individuals is itself an individual.

Roughly speaking, the Division Thesis says that there is nothing else except individuals and classes. But that is not exactly right. If we thought that Reality divided exhaustively into animal, vegetable, and mineral, that would not mean that there was no such thing as a salt beef sandwich. The sandwich is no counterexample, because the sandwich itself divides: the beef is animal, the bread is vegetable, and the salt is mineral. Likewise, the Division Thesis permits there to be a mixed thing which is neither an individual nor a class, so long as it divides exhaustively into individuals and classes. I accept a principle of *Unrestricted Composition*: whenever there are some things, no matter how many or how unrelated or how disparate in character they may be, they have a mereological fusion. (See sections 3.4 and 3.5 for further discussion.) That means that if I accept individuals and I accept classes, I have to accept mereological fusions of individuals and classes. Like the mereological fusion of the front half of a trout plus the back half of a turkey, which is neither fish nor fowl, these things can be mostly ignored. They can be left out of the domains of all but our most unrestricted quantifying. They resist concise classification: all we can say is that the salt beef sandwich is part animal, part vegetable, part mineral; the trout-turkey is part fish and part fowl; and the mereological fusion of Possum plus the class of all cat-whiskers is part individual and part class. Likewise, Reality itself – the mereological fusion of everything – is mixed. It is neither individual nor class, but it divides exhaustively into individuals and classes. Indeed, it divides into one part which is the most inclusive individual and another which is the most inclusive class.

If we accept the mixed fusions of individuals and classes, must we also posit some previously ignored classes that have these mixed fusions as members? No; we can hold the mixed fusions to be ineligible for membership. Mixed fusions are forced upon us by the principle of Unrestricted Composition. Classes containing them are not likewise forced upon us by a corresponding principle of unrestricted class-formation. That principle is doomed in any case: we dare not say that whenever there are some things, there is a class of them, because there can be no class of all non-self-members. Nor are classes containing mixed fusions forced upon us in any other way. Let us indulge our offhand reluctance to believe in them.

All I can say to defend the Division Thesis, and it is weak, is that as yet we have no idea of any third sort of thing that is neither individual nor class nor mixture of the two. Remember what an individual is: not necessarily a commonplace individual like Magpie or Possum, or a quark, or a space-time point, but anything whatever that has no members but is a member. If you believe in some remarkable non-classes – universals, tropes, abstract simple states of affairs, God, or what you will – it makes no difference. They're still individuals,

Are there any other Parts of Classes?

however remarkable, so long as they're members of classes and not themselves classes. Maybe the mixed fusions are disqualified from membership. Maybe some classes are 'proper classes' and are disqualified from membership. But rejecting the Division Thesis means positing some new and hitherto unheard-of disqualification, applicable this time to things that neither are classes nor have classes as parts, that can make things ineligible to be members. I wouldn't object to such a novel proposal, if there were some good theoretical reason for it. But so long as we have no good reason to innovate, let conservatism rule.

The Priority Thesis and the Fusion Thesis reflect our vague notion that somehow the individuals are 'basic' and 'self-contained' and that the classes are somehow a 'superstructure'; 'first' we have individuals and the classes come 'later'. (In some sense. But it's not that God made the individuals on the first day and the classes not until the second.) Indeed, these two theses may be all the sense that we can extract from that notion. We don't know what classes are made of – that's what we want to figure out. But we do know what individuals are made of: they're made of various smaller individuals, and nothing else.

From the First Thesis, the Division Thesis, the Priority Thesis, and the Fusion Thesis, our Second Thesis follows.

Proof Suppose the Second Thesis false: some class x has a part y that is not a class. If y is an individual, x has an individual as part; if y is a mixed fusion of an individual and a class, then again x has an individual as part; and by the Division Thesis those are the only possibilities. Let z be the fusion of all individuals that are part of x. Then z is an individual, by the Fusion Thesis. Now consider the difference $x - z$, the residue that remains of x after z is removed. (It is the fusion of all parts of x that do not overlap z.) Then $x - z$ has no individuals as parts, so it is not an individual or a mixed fusion. By the Division Thesis, it must be a class. We now have that x is the fusion of class $x-z$ with an individual z. Since $x-z$ is part of x, and not the whole of x (else there wouldn't have been any z to remove), we have that the class $x-z$ is a proper part of the class x. So, by the First Thesis, $x-z$ must be a proper subclass of x. Then we have v, a member of x but not of $x-z$. According to standard set theory, we then have u, the class with v as its only member. By the First Thesis, u is part of x but not of $x-z$; by the Priority Thesis, u is not part of z; so u has some proper part w that does not overlap z. No individual is part of w; so by the Division Thesis, w is a class. By the First Thesis, w is a proper subclass of u. But u, being one-membered, has no proper subclass. This completes a *reductio*.   QED

1.4 *The Null Set*

A consequence of our Second Thesis is that classes do not have the null set as part. Because it was a memberless member, we counted it as an individual, not a class; therefore it falls under our denial that individuals ever are parts of classes. To be sure, the null set is included in any class x: all its members – all none of them – are among x's members. But it never can be a *sub*class if it is not even a class. Were we hasty? Should we amend the Second Thesis, and the premises whence we derived it, to let the null set be a part of classes after all? I think not.

Some mereologists[5] posit a 'null individual' meant to play

---

[5] For one, R. M. Martin, 'A Homogeneous System for Formal Logic', *Journal of Symbolic Logic*, 8 (1943), pp. 1–23, especially p. 3; and 'Of Time and the Null Individual', *Journal of Philosophy*, 62 (1965), pp. 723–35. For another, Bunt, *Mass Terms and Model-Theoretic Semantics*, pp. 56–7.

The Null Set

a role in mereology that corresponds to the role of the null set in set theory. The null individual is supposed to be part of everything whatever. So if we take the fusion of the null individual and something else, we just get the something else back again, because the null individual was part of it already. And even if two things don't overlap in any ordinary way, they still have the null individual as a common part. And it has no proper parts; for if x is part of the null individual, the null individual also is part of x, wherefore x and the null individual are identical. If we accepted the null individual, no doubt we would identify the null set with it, and so conclude that the null set is part of every class.

But it is wellnigh unintelligible how anything could behave as the null individual is said to behave. It is a very queer thing indeed, and we have no good reason to believe in it. Such streamlining as it offers in formulating mereology — namely, that intersections of things come out well-defined even when they shouldn't — can well be done without. Therefore, reject the null individual; look elsewhere for the null set.

Should we perhaps reject the null set as well? Is it another misguided posit, meant to streamline the formulation of set theory by behaving in peculiar ways? I think not. It was misguided, perhaps, to call it a 'set'; and I have half-remedied that by declining at least to call it a 'class'. But the null set's behaviour is not, after all, so very peculiar. It is included in every class just because it lacks members — and lacking members is not so queer, all individuals do it. Also, the null set is more use than the null individual would be, and some of its services are less easily done without.

The null set serves, first, as a denotation of last resort for class-terms that fail to denote classes. It is the set-theoretical intersection of x and y, where x and y have no members in common (just as the null individual was invented to be the

Taking Classes Apart

mereological intersection of x and y when x and y didn't really overlap). It is the class of all self-members, since there are no self-members; it is the class of real numbers x such that $x^2 + 1 = 0$. This service is a mere convenience. It would be better to do without it than to purchase it by believing in some queer entity.

But, second, the null set also serves as an element of last resort. Suppose you repudiate the null set, and make do entirely with classes generated from some ordinary individual – let it be cat Possum. And suppose you reduce mathematical objects to classes, as usual – to do that, it matters not whether you begin with the null set or with Possum. Instead of the pure sets, you will have Possum himself; you will have Possum's *singleton*, or *unit class*, that contains him as its sole member; you will have the class of Possum's singleton and Possum himself; you will have the new singleton of that class; you will have the class of that new singleton and the old singleton and Possum; and so on *ad infinitum* and far beyond, until you have enough modelling clay to make the whole of mathematics. But now your confidence in the existence of the entire mathematical realm – all those numbers, matrices, curves, homeomorphisms, the lot – rests on your confidence that Possum exists! If I was fibbing when I said I had a cat of that name, shall mathematics fall? And where was mathematics before Possum was born to redeem it? If mathematics is to be safe, it had better rest on a surer foundation. Possum is not a sure thing. The null set is. You'd *better* believe in it, and with the utmost confidence; for then you can believe with equal confidence in its singleton, the class of that singleton and the null set, the new singleton of that class, the class of that new singleton and the old singleton and the null set, and so on until you have enough modelling clay to make the whole of mathematics.

(The null individual could do us no corresponding service.

The Null Set

If we have the null set, we have more-than-countless other things besides. If we had the null individual, we would have the null individual and that would be that. Mereology does not spin extravagant realms of being out of just one single thing.)

Must we then accept the null set as a most extraordinary individual, a little speck of sheer nothingness, a sort of black hole in the fabric of Reality itself?

Not really. We needn't be ontologically serious about the null set. It is useful to have a name that is guaranteed to denote some individual, but it needn't be a special individual with a whiff of nothingness about it. Ordinary individuals suffice. In fact, *any* individual will do for the null set — even Possum. Like any individual, he has the main qualification for the job — memberlessness. As for the second qualification, guaranteed existence, that is not really a qualification of the job-holder itself, rather it is a requirement on our method of selection. To guarantee that we will select some existing individual to be the null set, we don't need to select something that's guaranteed to exist. It is enough to make sure that we select from among whatever things happen to exist. We can select Possum, contingent though he be, so long as our method of selection would have found something else had Possum not been there to find.

(And what if there had existed no ordinary individuals whatsoever? — In that case, maybe we can let mathematics fall. Just how much security do we really need?)[6]

---

[6] I myself hold a thesis of plurality of worlds according to which (1) Possum could not have failed to exist altogether, though he might have been off in some other world and not a worldmate of ours; and (2) it is necessary that there be something — some individual — and not rather nothing. But I shall not rely on this thesis here, for fear that not all of you accept it.

The choice of an individual to serve as the null set is arbitrary. An arbitrary choice might be left unmade. We could say that what's true is what's true on all ways of making the choice; what's false is what's false on all ways; and what's true on some and false on others could be left indeterminate as to truth.[7] If I left the choice unmade, 'null set' would be harmlessly ambiguous: Possum would be not unequivocally not the null set, likewise Magpie, and likewise all other individuals.

But I prefer not to leave the arbitrary choice unmade. Instead, I make it — arbitrarily! — as follows. Instead of the unhelpful definition of the null set as the set without members, I adopt this

Redefinition: The *null set* is the fusion of all individuals.

That's an easy selection to specify; and it's guaranteed to select an individual to be the null set if there exist any individuals at all. It also introduces a handy name for one of the main subdivisions of Reality. For what it's worth, it respects our 'intuition' that the null set is no place in particular, no more one place than another. It's as far as can be from the notion that the null set is a speck of nothingness, and that's all to the good.

It may be that this choice of the null set offends against some 'intuition' you had that the null set was small and nondescript. If you think I am offering you a substitute for the null set as you have conceived of it hitherto, no harm done. We may agree to disagree, and proceed. But for myself, I think such an 'intuition' has no ground or authority, and the

---

[7] Here I apply Bas van Fraassen's method of 'supervaluations'. See van Fraassen, 'Singular Terms, Truth-Value Gaps and Free Logic', *Journal of Philosophy*, 63 (1966), pp. 481–95.

Consequences of the Main Thesis

fusion of all individuals is perfectly suitable to serve as the null set.

If the null set is the fusion of all individuals, then, by the Division, Priority, and Fusion Theses, the individuals are all and only the parts of the null set. And the urelements, individuals other than the null set, are the proper part of the null set.

1.5 Consequences of the Main Thesis

Our Main Thesis says that the parts of a class are all and only its subclasses. This applies, in particular, to one-membered classes: unit classes, or singletons. Possum's singleton has Possum as its sole member. It has no subclasses except itself. Therefore it is a mereological *atom*: it has no parts except itself, no proper parts. Likewise the singleton of Possum's singleton is an atom; and likewise for any other singleton.

Anything that can be a member of a class has a singleton: every individual has a singleton, and so does every set. The only things that lack singletons are the proper classes — classes that are not members of anything, and *a fortiori* not members of singletons — and those mixed things that are part individual and part class. And, of course, nothing has two singletons. So the singletons correspond one-one with the individuals and sets.

A class has its singleton subclasses as atomic parts, one for each of its members. It usually has other parts that are not singletons, namely subclasses with more than one member. However, if x is part of a class y, then x must have one or more singletons as parts, else x could have no members, could not be a class, and so could not be a subclass of y. In fact x must consist entirely of singletons, else the rest of x would be a part of y with no singletons as parts, so a part of a class that is not a subclass. A class is the union, and hence the fusion, of the singletons of its members. For example, the class of the two cats Possum and Magpie is the fusion of Possum's singleton and Magpie's singleton. The class whose three members are Possum, Magpie's singleton, and the aforementioned class is the fusion of Possum's singleton, Magpie's singleton's singleton, and the singleton of the aforementioned class. And so it goes.

If we take the notion of a singleton henceforth as primitive, we have this new definition of a class.

> *Redefinition*: A *class* is any fusion of singletons.

The members of a class are exactly those things whose singletons are subclasses of it; and so, by our Main Thesis, are exactly those things whose singletons are parts of it. An individual has no members and, by our Priority Thesis, it has no singletons as parts. The remaining case, by our Division Thesis, is a mixed fusion, part individual and part class. It does have singletons as parts, yet probably we would not want to say that it has members. So when we define membership (formerly primitive) in terms of the notion of singleton, we must write in a restriction to classes.[8]

> *Definition*: x is a *member* of y iff y is a class and the singleton of x is part of y.

By the Division and Priority Theses, something is an individual iff it has no class as part; and it has a fusion of

---

[8] Bunt does without this restriction, and accordingly grants that his concept of membership differs from that in set theory. See Bunt, *Mass Terms and Model-Theoretic Semantics*, p. 61.

singletons as part iff it has one or more singletons as parts. So we have another new definition.

> *Redefinition*: An *individual* is anything that has no singletons as parts.

Equivalently, since singletons are atoms, an individual is anything that overlaps no singletons; equivalently, since classes are fusions of singletons, an individual is anything that overlaps no classes. Hence the null set, as we have arbitrarily chosen it, is the fusion of all things that do not overlap the fusion of all singletons; which is to say that it is the mereological difference, Reality minus that fusion. It is what's left of Reality after all the singletons are removed.

1.6 More Redefinitions

Once we redefine membership and null set, we can proceed as usual to all other notions of set theory. But sometimes it is instructive not to go the long way round via membership, but rather to define other notions directly. Unfortunately, the special case of the null set keeps complicating the story.

Inclusion. The main case is that a class includes a subclass by having it as a part. But also a class includes the null set, which is not one of its subclasses and not one of its parts. The null set is not a class, but it too includes the null set. And although a class, or the null set, may be part of a mixed fusion, we probably would not want to call that a case of inclusion. So we have:

> *Redefinition*: y *includes* x iff (1) x is the null set and y is the null set or a class, or (2) x and y are classes and x is part of y.

Taking Classes Apart

Union. The main case is that a union of classes is their fusion. But the null set also may enter into unions; and other individuals, or mixed fusions, may not. So we have this:

Redefinition: The *union* of one or more things is defined iff each of them is either a class or the null set; it is the null set if each of them is the null set, otherwise it is the fusion of those of them that are classes.

1.7 Sethood and Proper Classes

A class is a set iff it is a member of something. If x is a member of something, it must have a singleton to be part of that thing, and if x has a singleton, the singleton is something that x is a member of. So a class is a set if it has a singleton, otherwise it is a proper class. But the null set also counts as a set, so sethood requires another disjunctive definition.

Redefinition: Something is a *set* iff either it is a class that has a singleton, or else it is the null set.

Redefinition: A *proper class* is a class that has no singleton.

For instance, the class of all sets that are non-self-members had better not be a set, on pain of Russell's paradox. Although it is indeed a non-self-member (everything is, according to the standard principle of *Fundierung*), that won't make it a self-member unless it is a set. So it isn't; it is a proper class, it has no singleton, and it cannot be a member of anything.

We dare not allow a *set* of all sets that are non-self-members, but there are two alternative ways to avoid it. One way would be to restrict composition: we have all the sets that are non-self-members and we have a singleton of each of these

A Map of Reality

sets, but somehow we have no fusion of all these singletons, so we have no class of all these sets. But there is no good independent reason to restrict composition. Mereology *per se* is unproblematic, and not to blame for the set-theoretical paradoxes; so it would be unduly drastic to stop the paradoxes by mutilating mereology, if there is any other remedy. (Just as it would be unduly drastic to solve problems in quantum physics by mutilating logic, or problems in the philosophy of mind and language by mutilating mathematics.) The better remedy, which I have adopted, is to restrict not composition, but rather the making of singletons. We *can* fuse all the singletons of sets that are non-self-members, thereby obtaining a proper class of sets, but this proper class does not in turn have a singleton. Unlike composition, the making of singletons is ill-understood to begin with, so we should not be surprised or disturbed to find that it needs restricting.

I do *not* say, note well, that we posit the proper classes because of their utility in formulating powerful systems of set theory. George Boolos has argued, and I agree, that we can get all the power we need by resorting to plural quantification.[9] And yet we have the proper classes willy-nilly, be they useful or be they useless. We do not go out of our way to posit them; rather we can't keep them away, given our Main Thesis and Unrestricted Composition.

1.8 A Map of Reality

Figure 1 is a map of all of Reality. The dots are mereological atoms. Those above the gap are singletons; those below are

---

[9] See Boolos, 'To Be is To Be the Value of A Variable (or To Be Some Values of Some Variables)', *Journal of Philosophy*, 81 (1984), pp. 430–49; 'Nominalist Platonism', *Philosophical Review*, 94 (1985), pp. 327–44.

Taking Classes Apart

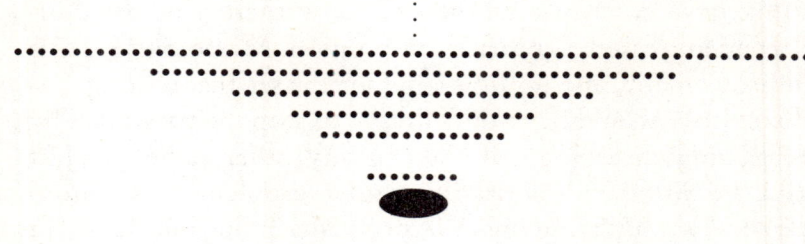

Figure 1

atomic individuals. The blob at the bottom is atomless gunk: an individual whose parts all have further proper parts. Anything made of the dots above the gap is a class. The biggest class is the thing made of all the dots above the gap. It is the universal class; every singleton is part of it, therefore everything is a member of it, save only those things that are not members of anything. It is a proper class, not a member of anything, and not a set; by the standard principle of Limitation of Size, it is too big to be a set. Many other classes likewise are too big to be sets. The smaller classes, however, are sets; and these classes have their own singletons, which are dots somewhere above the gap. A set is a member of just those classes that have its singleton as a part; and by the standard principle of *Fundierung*, it can never itself be one of the classes that its singleton is part of. To make sure of this, let a set's singleton always go just above the set itself. (That's why there's no top, and that's why there are more and more dots as we ascend.) The smallest classes are the singletons, each one being just one dot, and of course these are all small enough to be sets.

Anything made of the dots below the gap, the blob of

gunk, or some of each is an individual. (Maybe there isn't any gunk; or maybe there aren't any atomic individuals; or maybe there are both, and that is the case shown.)[10] The biggest one is the null set, the fusion of all individuals; it is made of all the atomic individuals plus all of the atomless gunk. Each individual has its singleton, somewhere up just above the gap, and is a member of all classes that have that singleton as a part. That is why the part of the map above the gap must be so much bigger: so that there will be enough singletons to go around.

Straddling the gap are the 'salt beef sandwiches': the mixed fusions, ignored but undenied, consisting partly of dots or gunk from below and partly of dots from above. These have no singletons and are not members of anything.

1.9 Nominalistic Set Theory Revisited

Years ago, I wrote a paper called 'Nominalistic Set Theory'.[11] By 'nominalistic' I meant 'setless', in the Harvard fashion; nominalism in the traditional sense of the word – repudiation of universals – had nothing to do with it. I argued that we could imitate set theory, to a limited extent, by mereological means.

One reason why the fusion of several things is different

---

[10] Bunt disagrees; he takes it as axiomatic that *every* atom is a singleton (*Mass Terms and Model-Theoretic Semantics*, p. 60). So his individuals, if there are any, must consist of gunk; for instance he cannot countenance space-time points (unless they are singletons). I find this assumption hasty. Later he reopens the question (p. 291), but only for purposes of a formal comparison of systems.

[11] David Lewis, 'Nominalistic Set Theory, *Noûs*, 4 (1970), pp. 225–40.

from the class of them is that a fusion does not, in general, have a unique decomposition into parts. However, it may happen that if we distinguish some parts of a fusion as *nice* parts, then a fusion will have a unique decomposition into nice parts.

Cats are nice. A fusion of cats has a unique decomposition into cats; the fusion of cats $c_1, c_2, c_3, \ldots$ is identical to the fusion of cats $d_1, d_2, d_3, \ldots$ only if each if the cs is one of the ds and each of the ds is one of the cs. That is because one cat is never a proper part of another, and one cat is never a proper part of the fusion of several others. More generally, it is because cats never overlap.

Maximal spatiotemporally connected parts are nice. For the sake of this example, let us assume that spatiotemporal things consist entirely of point-sized particulars; ignore universals, if such there be. Suppose we have some notion of what it means for two things to *touch* spatiotemporally: roughly, that points of one are arbitrarily close to points of the other. Then something x is *connected* iff there are no y and z such that x is the fusion of y and z and y does not touch z; x is a *maximal connected part* of w iff x is a connected part of w that is not a proper part of any other connected part of w. No two maximal connected parts of a fusion overlap, so a fusion has a unique decomposition into its maximal connected parts.

Mereological atoms are nice. No two atoms overlap, so any fusion of atoms has a unique decomposition into its atoms.

Fix on some specific definition of 'nice part'. Then fusions of nice parts imitate classes, and the relation 'nice part of' imitates the membership relation, in this way: whenever we have some suitable things, we will have a fusion that has all and only those things as nice parts. Say it like this: we have a pseudo-class that has all and only those things as its pseudo-members.

Nominalistic Set Theory Revisited

This first stab at setless theory doesn't go far, because the only things that can qualify as pseudo-members of pseudo-classes are the things that will be nice parts. If nice parts are maximal connected parts, for instance, then only a connected thing can ever be a pseudo-member of a pseudo-class. If we try to put in a disconnected thing, we cannot get that thing itself back as a pseudo-member. The closest we can come is to get back its several maximal connected parts. That's the best case. Sometimes other unwelcome things happen instead.

To improve matters, we resort to coding. Define some scheme of coding on which things that are not themselves nice may nevertheless have nice codes. Suppose we can encode the things we want to have as the pseudo-members of our pseudo-classes. Then we can say that x is a *pseudo-member* of w, relative to a given definition of 'nice part' and a given scheme of coding, iff some nice part of w is a code for x. In 'Nominalistic Set Theory', I chose maximal connected parts as my 'nice parts' and tried various geometrically definable codings. (The 'first stab' was the case where I chose the coding to be identity.) I got imitations of set theory, but they worked only under especially favourable conditions.

Example. Suppose the things of interest to us are made out of points in an infinite grid. Each point in the grid is next to eight others: two vertical neighbours, two horizontal, four diagonal. Two things *touch* iff some point of one is next to some point of the other. Nice parts are maximal connected parts. The *interior* of something consists of those points of it which are not next to any points that are not parts of it. Something x is a *code* for something y iff y is the interior of x. If we have disconnected things, we can make connected codes for them by wrapping them in skins and running strings from one part to the other. Thus: suppose we have two disconnected things a and b, each composed of the points

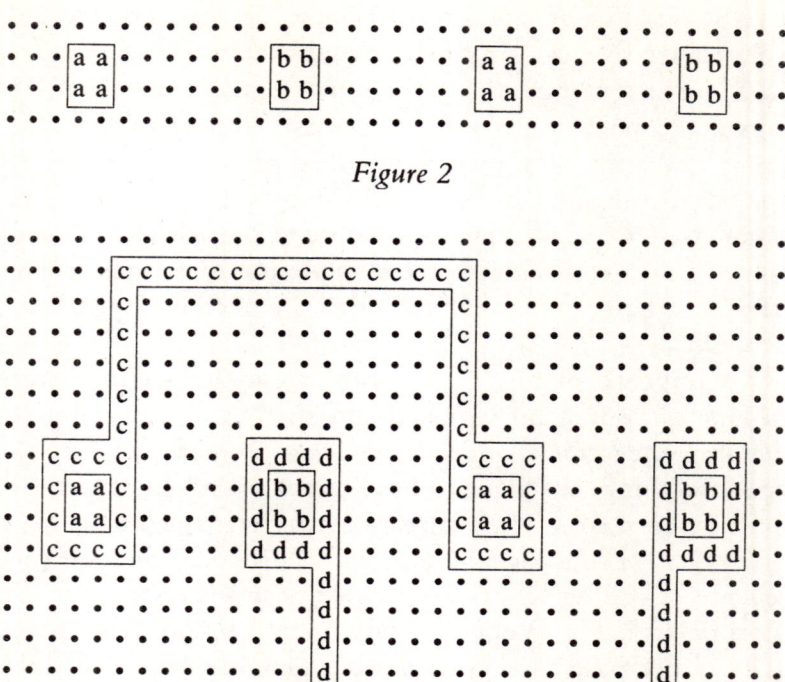

Figure 2

Figure 3

so labelled as shown in figure 2. Then we can form a pseudo-class as shown in figure 3. The fusion $a+c$ is a connected code for a, the fusion $b+d$ is a connected code for b, these two codes are the two nice parts of the fusion $a+c+b+d$, wherefore that fusion is a pseudo-class of a and b. The coding is redundant: the strings could have taken different routes. So we cannot speak of *the* pseudo-class of a and b. *Any* fusion whose nice parts are codes for a and for b is a pseudo-class of a and b.

Besides the lack of a unique decomposition, another reason

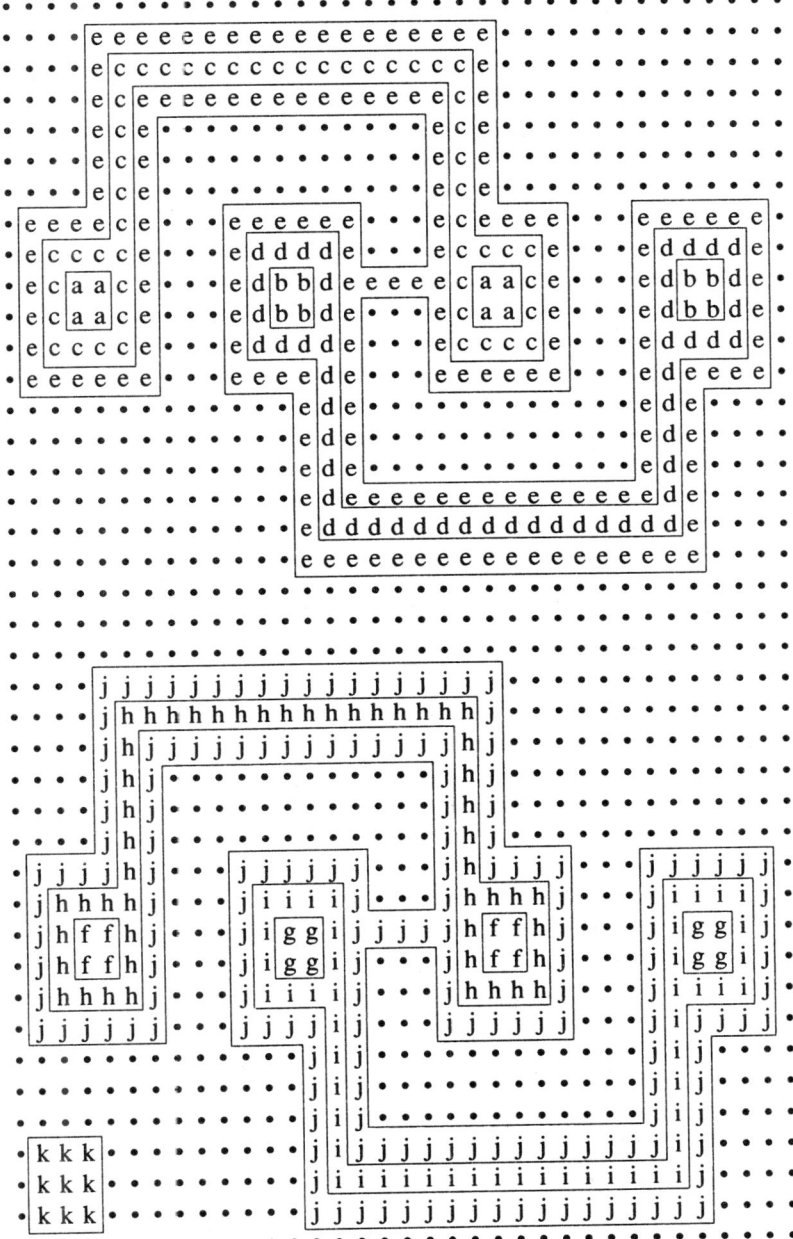

Figure 4

Taking Classes Apart

why the fusion of several things differs from the class of them is that fusion offers nothing to correspond to the set-theoretical hierarchy of individuals, classes of individuals, classes of classes of individuals (or mixed classes, containing classes of individuals and also individuals), and so on, *ad infinitum* and beyond. But if coding can be iterated, then fusion plus coding do offer at least the beginning of a hierarchy. We saw that $a+c+b+d$ – call it P – was a pseudo-class of *a* and *b*. By attaching a skin and strings, as shown in figure 4, we can make $a+c+b+d+e$, a code for P. Likewise $f+h+g+i$ – call it Q – is a pseudo-class of *f* and *g* and $f+h+g+i+j$ is a code for Q. Now $a+c+b+d+e+f+h+g+i+j$, which has these two codes as its nice parts, is the pseudo-class of P and Q. And we could go on a little longer. There's a problem: when P turns up as a pseudo-member of the pseudo-class of P and Q, how can we tell that it's meant to be taken as a code for a pseudo-class, and further broken up into nice parts and decoded? How could we tell whether k, if it turned up as a pseudo-member of some other pseudo-class, was meant to be further decoded? It could be: it is a code for its central point.

It is easy to think of more problems. What if the result of going up the hierarchy is that our grid gets too crowded to make room for any more strings and skins? (How much would it help to work in more than two dimensions?) What if we want to make a pseudo-class of things that touch? Or almost touch? Or worse, overlap? What if we want to make a pseudo-class of a disconnected thing and something that completely surrounds one of its parts? You can define ever more complicated geometrical codings to cover ever more difficult cases. You can while away many an hour. But you cannot hope to arrive at a mereological-cum-geometrical reconstruction of the entire realm of mathematical objects. To see this, it is enough to remember that, out of a countable

Nominalistic Set Theory Revisited

grid of points, there are only continuum many different fusions; but the mathematical realm is much bigger than that.

But imagine now that we could have a coding for free, and never have to to define it. Farewell honest toil! What conditions would we like it to meet?

We'd like to avoid crowding; so let the codes be small. In fact, let them be atoms; because whenever we have a fusion of atoms, its atoms are nice parts of it. (Something that is a maximal connected part of one fusion may not be a maximal connected part of another; its niceness is relative to the fusion. Not so for the niceness of atoms.)

We must avoid ambiguity. We'd also like to avoid redundancy, in order to use our supply of codes efficiently. So let the encoded things and their codes correspond one-one.

We'd like to avoid any confusion between our original individuals and our codes, or fusions of codes. So let the codes be entirely distinct from the individuals.

We'd like to be able to iterate coding and build up a hierarchy. So let the codes, and the fusions of codes, have their own codes in turn — so far as possible. But there must be a limitation. Not all code-fusions can have codes.

Proof We apply the reasoning of Cantor's Theorem and Russell's Paradox. Call a code-fusion *normal* iff it has a code and its code is not part of it. By Unrestricted Composition, we have the fusion of all codes of normal code-fusions; call it n. Has n a code? Then by definition of n, n's code is part of n iff n is normal. By definition of normality, n's code is part of n iff n is not normal. We conclude that n is a code-fusion that has no code.   QED

In fact, if we are not to run out of codes, the great majority of code-fusions must remain codeless. One way to ration a

Taking Classes Apart

limited supply of codes is to allocate them only to the comparatively small code-fusions. Then the great majority of code-fusions would go without, because the great majority of code-fusions are not small. To take a miniature example: if we had countably many codes, then each of the countably many finite code-fusions could have one, leaving the continuum many infinite code-fusions uncoded. To be sure, this principle of Limitation of Size is not the only possible way of selecting a favoured few code-fusions to receive codes. Alternative principles of rationing might allow codes to some few of the infinite fusions, or deny codes to some or all of the finite ones. But standard iterative set theory uses Limitation of Size, and that is what our setless imitation is meant to imitate.

Code-fusions are pseudo-classes. Those with codes of their own are pseudo-sets, the rest are proper pseudo-classes. Also there is the null pseudo-set, something or other that has no codes as parts and therefore has no pseudo-members. Whenever something has a code, that code is a pseudo-singleton.

If only heaven would grant us an ideal coding, one that satisfied all the desiderata just listed, and one that provided codes for all manner of ordinary individuals, then the program of setless set theory would succeed. The pseudo-classes made by mereological fusion of codes would be every bit as good as the real thing.

I say they would *be* the real thing. If setless set theory works that well, its classes are no longer pseudo. There is nothing we know about the real classes that distinguishes them from the 'imitations'. Somehow, I know not how, heaven *has* granted us an ideal coding. We call the codes 'singletons'. Classes *are* code-fusions, with singletons as their nice – that is, atomic – parts. Something is a member of a class iff some nice part of that class is a code for that thing.

2

The Trouble with Classes

2.1 Mysterious Singletons

Cantor taught that a set is a 'many, which can be thought of as one, i.e., a totality of definite elements that can be combined into a whole by a law'. To this day, when a student is first introduced to set theory, he is apt to be told something similar. 'A set is a collection of objects.... [It] is formed by gathering together certain objects to form a single object' (Shoenfield). A set or class is 'constituted by objects thought of together' (Kleene). 'Roughly speaking, a set is a collection of objects and is thought to have an independent existence of its own...' (Robbin). Maybe also, or instead, he will be given some familiar examples: '...for example, the set of all even numbers is considered to be just as real as any particular even number such as 2 or 16' (Robbin); 'A pack of wolves, a bunch of grapes, or a flock of pigeons are all examples of sets of things' (Halmos).[1]

---

[1] Georg Cantor, *Gesammelte Abhandlungen mathematischen und philosophischen Inhalts*, ed. Ernst Zermelo (Springer, 1932), p. 204; Joseph R.

The Trouble with Classes

But after a time, the unfortunate student is told that some classes — the singletons — have only a single member. Here is a just cause for student protest, if ever there was one. This time, he has no 'many'. He has no element*s* or object*s* — I stress the plural — to be 'combined' or 'collected' or 'gathered together' into one, or to be 'thought of together as one'. Rather, he has just one single thing, the element, and he has another single thing, the singleton, and nothing he was told gives him the slightest guidance about what that one thing has to do with the other. Nor did any of those familiar examples concern single-membered sets. His introductory lesson just does not apply.

(He might think: whatever it is that you do, in action or in thought, to make several things into a class, just do the same to a single thing and you make it into a singleton. Daniel Isaacson suggests this analogy: How do you make several paintings into an art collection? Maybe you make a plan, you buy the paintings, you hang them in a special room, you even publish a catalogue. If you do the same, but your money runs out after you buy the first painting on your list, you have a collection that consists of a single painting. — But this thought is worse than useless. For all those allusions to human activity in the forming of classes are a bum steer. Sooner or later our student will hear that there are countless classes, most of them infinite and miscellaneous, so that the vast majority of them must have somehow got 'formed' with absolutely no attention or assistance from us. Maybe we've formed a general concept of classes, or a theory of them, or some sort of sketchy mental map of the whole of

---

Shoenfield, *Mathematical Logic* (Addison-Wesley, 1967), p. 238; Stephen C. Kleene, *Mathematical Logic* (John Wiley, 1967), p. 135; Joel W. Robbin, *Mathematical Logic: A First Course* (W. A. Benjamin, 1969), p. 171; Paul R. Halmos, *Naive Set Theory* (Van Nostrand, 1960), p. 1.

Mysterious Singletons

set-theoretical Reality. Maybe we've formed a few mental representations of a few very special classes. But there just cannot be anything that we've done to all the classes one at a time. The job is far too big for us. Must set theory rest on theology?)

We were told nothing about the nature of the singletons, and nothing about the nature of their relation to their elements. That might not be quite so bad if the singletons were a very special case. At least we'd know about the rest of the classes. But since all classes are fusions of singletons, and nothing over and above the singletons they're made of, our utter ignorance about the nature of the singletons amounts to utter ignorance about the nature of classes generally. We understand how bigger classes are composed of their singleton atoms. That's the easy part: just mereology. *That's* where we get the many into one, the combining or collecting or gathering. Those introductory remarks (apart from the misguided allusions to human activity) introduced us only to the *mereology* in set theory. But as to what is distinctively set-theoretical – the singletons that are the building blocks of all classes – they were entirely silent. Dr Jekyll was there to welcome us. Mr Hyde kept hidden. What do we know about singletons when we know only that they are atoms, and wholly distinct from the familiar individuals? What do we know about other classes, when we know only that they are composed of these atoms about which we know next to nothing?

Set theory has its unofficial axioms, traditional remarks about the nature of classes. They are never argued, but are passed along heedlessly from one author to another. One of these unofficial axioms says that the classes are nowhere: they are outside of space and time. But why do we think this? Perhaps because, wherever we go, we never see them or stumble over them. But maybe they are invisible and intangible.

Maybe they can share their locations with other things. Maybe they are somewhere. Where? If Possum's singleton were elsewhere than Possum himself, it would presumably be an obnoxiously arbitrary matter where it was. If it's in Footscray, why there instead of Burwood? But perhaps Possum's singleton is just where Possum is. Perhaps, indeed, every singleton is just where its member is. Since members of singletons occupy extended spatiotemporal regions, and singletons are atoms, that would have to mean that something can occupy an extended region otherwise than by having different parts that occupy different parts of the region, and that would certainly be peculiar. But not more peculiar, I think, than being nowhere at all – we get a choice of equal evils, and cannot reject either hypothesis by pointing to the repugnancy of the other. If every singleton was where its member was, then, in general, classes would be where their members were. The class of Magpie and Possum would be divided: the part of it that is Magpie's singleton would ocupy the region where Magpie is, the part that is Possum's singleton would occupy the region where Possum is, and so the entire class would occupy the entire spatiotemporal region where Magpie and Possum are.

Perhaps when we say that classes are outside of space and time, it is especially the 'pure' classes that we have in mind. These are the classes built up from the null set and nothing else: the null set, its singleton, its singleton's singleton, the class of the null set and its singleton, and so on. Maybe we think that these classes, at any rate, are nowhere; because we start by thinking that classes are where their members are, we conclude that a class with no members must be nowhere, and finally we conclude that all the pure classes must share the location or unlocation of the null set that is their common ancestor. But even if genuine classes are where their members are, the null set is no genuine class, only a 'set' by courtesy. If

Mysterious Singletons

the null set is the fusion of all individuals, as I arbitrarily stipulated, then it is everywhere, rather than nowhere. If the pure classes share its location, they too are everywhere — nowhere *in particular*, to be sure, but by no means outside of space and time.

I don't say the classes are in space and time. I don't say they aren't. I say we're in the sad fix that we haven't a clue whether they are or whether they aren't. We go much too fast from not knowing whether they are to thinking we know they are not, just as the conjurer's dupes go too fast from not seeing the stooge's head to thinking they see that the stooge is headless.

Another unofficial axiom says that classes have nothing much by way of intrinsic character. We know that's not quite right: to be an atom, or to be a fusion of atoms, or to be a fusion of exactly seventeen atoms, are matters of intrinsic character. However, these are not matters that distinguish one singleton from another, or one seventeen-membered class (a seventeenfold fusion of singletons) from another. Are all singletons exact intrinsic duplicates? Or do they sometimes, or do they always, differ in their intrinsic character? If they do, do those differences in any way reflect differences between the character of their members? Do they involve any of the same qualities that distinguish individuals from one another? Again we cannot argue the case one way or the other, and if we think we know that classes have no distinctive intrinsic character, probably that's like thinking we know that the stooge is headless.[2]

---

[2] Some philosophers propose that things have their qualities by having them as parts. The qualities might be repeatable universals, as in the main system of Nelson Goodman, *The Structure of Appearance* (Harvard University Press, 1951); or they might themselves be particular, as in Donald C. Williams, 'On the Elements of Being' in *Principles of Empirical Realism* (Charles Thomas, 1966), and Keith Campbell, *Abstract Particulars*

The Trouble with Classes

Sometimes our offhand opinions about the nature of classes don't even agree with one another. When Nelson Goodman finds the notion of classes 'essentially incomprehensible' and says that he 'will not willingly use apparatus that peoples his world with a host of ethereal, platonic, pseudo entities',[3] we should ask which are they: ethereal or platonic? The ether is everywhere, and one bit of it is pretty much like another; whereas the forms are nowhere, and each of them is unique. Ethereal entities are 'light, airy or tenuous', says the dictionary, whereas the forms are changeless and most fully real (whatever that means). If we knew better whether the classes were more fittingly called 'ethereal' or 'platonic', that would be no small advance!

Because we know so little about the singletons, we are ill placed even to begin to understand the relation of a thing to its singleton. We know what to call it, of course – membership – but that is all. Is it an external relation, like the spatiotemporal relations of distance? Or is it an internal relation, like a relation of intrinsic similarity or difference? Or is it a combination of the two? Or something else altogether? It cannot be entirely an internal relation: when two individuals are exact intrinsic duplicates, as sometimes happens, they cannot differ in their internal relations to any third thing, yet something is the singleton of one but not the other. If classes are outside space and time, then membership cannot be at all a matter of spatiotemporal relations between the thing and the singleton. Even if classes are where their members are, it cannot be entirely a matter of spatiotemporal relations: Possum and his singleton occupy the same region on this hypothesis,

---

(Blackwell, 1990). A mereologically atomic singleton could have no qualities as proper parts. Still it would not follow that a singleton had no qualitative character. It might *be* a quality.

[3] Goodman, *The Structure of Appearance*, section II. 2.

Van Inwagen's Tu Quoque

and so are just alike in their spatiotemporal relation to any third thing, yet something is the singleton of one and not the other. (Nor will it help to add an extra quasi-spatial dimension of 'rank' in the set-theoretical hierarchy – the vertical in our map of Reality – because the singleton of the singleton of the singleton of Possum and the singleton of the class of Possum and his singleton will be at the same rank, as well as the same spatiotemporal location.)

We cannot use the internal structure of a singleton to encode the information which thing it has as its member; because a singleton, being an atom, has no internal structure. Nor can we use familiar qualities that the singleton might conceivably share with familiar individuals. There are different singletons for all the individuals and all the sets. There just aren't enough familiar qualities to encode that much information. Even if we suppose that singletons have spatiotemporal locations to help carry the information, we still fall very far short. It seems that we have no alternative but to suppose that the relation of singleton to member holds in virtue of qualities or external relations of which we have no conception whatsoever. Yet we think we do somehow understand what it means for a singleton to have a member!

Finally, it's no good saying that a singleton has x as its member because it has the property: being the singleton of x. That's just to go in a circle. We've named a property; but all we know about the property that bears this name is that it's the property, we know not what, that distinguishes the singleton of x from all other singletons.

2.2 Van Inwagen's Tu Quoque

Elsewhere I have complained about the difficulty of understanding the relation that allegedly relates concrete things to the abstract simple possibilities – propositions or properties –

that they realize. Peter van Inwagen replied with a *tu quoque*: I accept set theory, yet the relation of member to set seems difficult to understand for exactly the same reasons.[4] Van Inwagen concludes that my complaints are as damaging to the set theory I accept (and to classical mathematics as well) as they are to the doctrine of abstract simple possibilities I reject. Why should they be lethal to the latter and not the former?

His point is very well taken, and my present complaints against membership (or rather, against the special case of the member-singleton relation) amount to my endorsement of it. But are the parallel complaints lethal, or are they not?

I suppose they're equally formidable in both cases, but not quite lethal in either case. It's a nasty predicament to claim that you somehow understand a primitive notion, although you have no idea how you could possibly understand it. That's the predicament I'm in when I accept the notion of singleton, and that's the predicament I claim others are in if they accept the alleged notion of realization of abstract simple possibilities. Either is bad, and both together are worse. What price escape? It seems (and van Inwagen agrees) that rejecting the notion of singleton means rejecting present-day set-theoretical mathematics. Hot though it is in the frying pan, that fire is worse. But in the case of the abstract simple possibilities, escape can be had by accepting the doctrine of plurality of worlds. That's much easier to understand, though not very easy to believe. This time, say I, the price is right – escape is worth it. The difference is almost entirely between the fires, not between the frying pans.

Almost entirely: I have to add that the *tu quoque* is imperfect. I posed a dilemma: is the relation of the concrete world

---

[4] For my complaint, see Lewis, *On the Plurality of Worlds*, pp. 176–91; for his reply, see Peter van Inwagen, 'Two Concepts of Possible Worlds', *Midwest Studies in Philosophy*, 11 (1986), pp. 185–213, especially pp. 207–10.

Van Inwagen's Tu Quoque

to the abstract possibilities internal or external? So far as the 'internal' horn goes, the *tu quoque* is complete. Not so for the 'external' horn – but there's enough left of the *tu quoque*, even there, to drive home how nasty it is to have to believe in the member-singleton relation.

Taking the 'external' horn, one big part of my complaint was that we had no independent grasp on what sort of relation we were dealing with. That applies equally to the case of the singletons. But another part of my complaint was that we get a mysterious necessary connection: if the concrete world has a certain intrinsic qualitative character, it *must* stand in the external relation to certain abstract simples and not others. Why? We would expect, as in the case of shape and distance, say, that intrinsic character and external relations can vary independently.

Can you turn a parallel complaint against the singletons? – Try it thus.[5] Suppose the relation of member to singleton is external. Why *must* Possum be a member of one singleton rather than another? Why isn't it contingent which singleton is his? – But to this I have a reply. On my theory of modality the question becomes: why doesn't some other-worldly counterpart of Possum have a singleton which isn't a counterpart of the singleton that Possum actually has? And my answer is: what makes one singleton a counterpart of another exactly is their having counterpart members. Or rather, I say that it's a flexible matter which things we count as counterparts; one way, the answer just given holds; other ways, it's not even true that a counterpart of Possum must have a counterpart of his singleton.[6]

---

[5] Van Inwagen, 'Two Concepts of Possible Worlds', p. 210.
[6] On counterparts, see David Lewis 'Counterpart Theory and Quantified Modal Logic', *Journal of Philosophy*, 65 (1968), pp. 113–26; *On the Plurality of Worlds*, passim.

The Trouble with Classes

Now, can the defender of abstract possibilities answer me in a parallel way? – No, because the two complaints involve different kinds of necessary connection. The complaint against singletons involves a necessary connection between external relations and the *identity* of the relatum; whereas the complaint against abstract simple possiblilities involves a necessary connection between external relations and the *qualitative character* of the relatum. The method of counterparts does not apply to the latter problem.

2.3 On Relations that Generate

I dislike classes because they burden us with mysteries. Nelson Goodman dislikes classes for a different reason. In 'A World of Individuals',[7] he advances a compelling principle: the 'generating relation' of a system should never generate two different things out of the very same material. There should be no difference without a difference in content. There has to be something to *make* the difference, and it has to be something built, directly or indirectly, into one thing but not the other. We can state this more precisely. Let R be the generating relation; if we start with some sort of non-transitive relation of direct or immediate generation, let R be not that but rather its ancestral. A *minimal element* is something to which nothing else bears R. The *content* of a thing consists of

---

[7] In *Philosophy of Mathematics: Selected Readings*, ed. Paul Benacerraf and Hilary Putnam (Prentice Hall, 1964); and in Nelson Goodman, *Problems and Projects* (Bobbs-Merrill, 1972). The paper was originally published in two parts: 'A World of Individuals' in I. M. Bochenski, Alonzo Church, and Nelson Goodman, *The Problem of Universals* (University of Notre Dame Press, 1956); and the important appendix as a separate note 'On Relations that Generate', *Philosophical Studies*, 9 (1958), pp. 65–6.

On Relations that Generate

those minimal elements that bear R to it (or of the thing itself, if it is already minimal). Then Goodman's principle requires that no two things have exactly the same content. Otherwise, R is not acceptable as a generating relation.

If we take the proper part relation as our generating relation, Goodman's principle is satisfied: it is a principle of mereology that no two things consist of exactly the same atoms. (Not if the two things consist entirely of atoms, anyway. The principle needs revision if we accept atomless gunk as a genuine possibility.) If we start with membership, on the other hand, and take its ancestral as our generating relation, and accept anything remotely resembling the normal existence axioms of set theory, then we find Goodman's principle violated in a big way. Countless things are generated out of exactly the same content. Possum's singleton, his singleton's singleton, the class of him and his singleton, and ever so many more things – all are generated out of Possum alone. There is difference galore without difference of content.

Roughly, I want to say that I accept Goodman's principle (ignoring gunk), but I deny that the ancestral of membership is truly a generating relation. But then I have some urgent explaining to do, because Goodman *stipulates* that the ancestral of membership *is* a generating relation in a system founded on set theory. What is a generating relation, anyway?

Goodman characterizes the idea in a preliminary way with a volley of near-synonyms. A generating relation is one whereby entities are 'generated', or 'comprised of', or 'made up of', or 'composed of' other entities; and its converse is a relation whereby one entity 'breaks down into' others. But then he proceeds to an example in which it is stipulated that the generating relations of two systems are in one case the proper part relation and in the other case the ancestral of membership – the challenge is to find out which is which. Finally (in an appendix originally published apart from the rest of the paper) he gives the official definition: 'the generating relation of a system is the proper part relation or the ancestral of ... membership or the logical sum of the two, as they occur in the system'. Given this stipulation, of course I cannot deny that the ancestral of membership is a generating relation; given the stipulation, I deny instead that Goodman's principle has any force. It says only that if a relation is either the proper part relation, or the ancestral of membership, or the logical sum of the two, then it must obey a certain condition which the proper part relation does obey and the ancestral of membership (or the sum of the two) does not. Why should anyone believe that, unless because he renounces classes already?

But if we go against Goodman's wishes and ignore his official stipulation, then we have an argument more worthy of attention. Suppose we thought that we had a broad notion of composition (or 'generation', or 'making up', or what you will) and that mereological composition was only one species of composition among others. Then we could very well think that every kind of composition should obey Goodman's principle, even if there were some sort of unmereological composition alongside the mereological sort. If so, we should conclude that generation of classes via the ancestral of membership is not – contrary perhaps to our first impressions – a legitimate sort of unmereological composition.

And with that I agree. But not because I think it illegitimate. Rather, because the generation of classes is not an unmereological sort of composition. In so far as it is unmereological, it isn't composition; in so far as it is composition, it is mereological. When something is a member of a class, first the thing is a member of its singleton – that part is not composition at all. Then its singleton is part of the class – that part is mereological.

Quine on Urelements and Singletons

When the thing is a member of its singleton, there isn't one thing made out of many; there is one thing made out of one thing. Further, the thing and its singleton are entirely distinct. (If member x and singleton y were not entirely distinct, then, since y is an atom, either x and y would be identical, or y would be a proper part of x and x would be a class, or y would be a proper part of x and x would be a fusion of a class and an individual. In the first and second cases, x would be a member of itself, which is impossible; in the third case, x could not be a member of anything.) What does this relation of one thing to another wholly distinct thing have to do with composition, in no matter how broad a sense? You might as well say that marrying someone is an odd sort of composition, in which your spouse is generated from you! For better or worse, the ancestral of membership is officially a generating relation. But if it's a generating relation in the (vague) sense given by Goodman's preliminary volley of near-synonyms, you might as well say that the ancestral of 'being married to someone who is part of' is a generating relation too. Apart from the fact that marriage is better understood, the cases are parallel.

2.4 Quine on Urelements and Singletons

Quine has discovered a nice way to remove all mystery about the nature of *some* singletons.[8] He reinterprets the predicate of membership when applied to urelements (for him, 'individuals', since he does not call the null set an individual) to mean what we would originally have called membership-or-identity. He thereby counts urelements as self-members; thereby he identifies each urelement with its singleton;

---

[8] See W. V. Quine, *Set Theory and Its Logic* (Harvard University Press, 1963), pp. 31–4.

indeed, he defines an urelement (an 'individual') as something identical to its singleton. If Possum is his own singleton, then also his singleton is his singleton's singleton, and so he himself is his singleton's singleton; likewise he is his singleton's singleton's singleton, and so on. So we have no more mystery about the nature of any of these singletons than we do about Possum himself.

Quine's plan cannot be extended, else we give most classes spurious members. If the null set were its own singleton, it would have itself for a member, though it was supposed to be memberless. If the class of Magpie and Possum were its own singleton, it would have itself as a third member along with Magpie and Possum. And so on. So if we followed Quine, we could at best remove only some of our mysteries about the nature of singletons and their relation to their members.

But in fact we cannot even begin to follow Quine. Cats are urelements, so for Quine they are their own singletons. The molecules of which cats are composed also are urelements, so again their own singletons. The class of cats, for us, is the fusion of cat-singletons; so if cats are their own singletons, that class is the fusion of cats. Likewise the class of cat-molecules, if these too are their own singletons, is the fusion of cat-molecules. The fusion of cats and the fusion of cat-molecules are identical. But the class of cats and the class of cat-molecules are not identical. Sad to say, Quine's idea is not for us. Nice as it is, I submit that our mereology of classes is nicer; and the two cannot coexist on pain of collapse.

2.5 The Lasso Hypothesis

We learnt in school to picture a class by drawing pictures of its members, and then drawing a picture of a lasso around

The Lasso Hypothesis

them. What if this were the exact truth of the matter? Maybe the singleton of something x is not, after all, an atom; but rather consists of x plus a lasso. That gives singletons an internal structure after all; and so might seem to shed light, in a speculative way, both on the nature of the singletons and on their relation to their members.

(Picture it a different way, if you'd rather: a singleton is its member *encapsulated*. What surrounds the member is not a soft loop but a hard shell. So tough is this shell that when we try to break a class down into its smallest parts, we do not quite succeed. We get the singleton subclasses, but these are not yet genuine mereological atoms. Rather they are the encapsulated members of the class, each one consisting of member plus shell, which have remained unsplit despite our best efforts. – This picture needs more than a grain of salt! For when something consists of parts, that is not because of anything we have done to break it into parts. It makes no difference whether we are able to separate the parts, or whether it even makes sense to talk of separating the parts. So ignore the breaking, ignore the failing to break, ignore the toughness of the shells; what's left is just the hypothesis that the singleton of x consists of x plus something else. Whether you call this something else a 'shell' or a 'lasso' is neither here nor there.)

The Lasso Hypothesis says that a singleton has proper parts, namely its lasso, its member, and its member's proper parts, if any. So it contradicts our Main Thesis, which says that a class has no parts but its subclasses, and which thereby implies that a singleton has no proper parts. So it must also contradict one of the premises whence the Main Thesis follows. I take it to contradict the Division Thesis: the lasso would be neither individual nor class nor mixed fusion of the two. But the Division Thesis could go, in a good cause; all I said in its favour was that we had no good reason to abandon it; and if abandoning it could buy us clarification of how a singleton is related to its member, that would be a good reason indeed!

But now ask: if the singleton of x consists of x plus a lasso, and the singleton of y consists of y plus a lasso, can it ever be that the same lasso is used twice over? (Can it even be that one single lasso will do for making all the singletons?) Or must it be a different lasso each time, one lasso per singleton? The answer is that it must be a different lasso each time.

Proof Say that lasso L *fits* x iff the singleton of x is the fusion $L+x$. Suppose for *reductio* that we have two different things x and y, and yet one lasso L fits both: x's singleton is $L+x$, y's singleton is $L+y$. Case 1: Either x and y are both individuals, or else they are both classes. If they're classes they're sets, and their union is a set too. Either way, another thing that has a singleton is the fusion $x+y$; its singleton is $M+(x+y)$. (We needn't ask whether M and L are the same lasso.) Now the class whose members are x and $x+y$ is the fusion of the two singletons, that is $(L+x)+(M+(x+y))$, which reduces to $L+M+x+y$; and likewise the class whose members are y and $x+y$ is $(L+y)+(M+(x+y))$, which also reduces to $L+M+x+y$. So the classes are identical. It follows that x and y are identical, contra our supposition. Case 2: One of x and y, let it be x, is an individual; the other, y, is a class. Then the class of x and y is $(L+x)+(L+y)$, which reduces to $L+x+y$; and the union of y with the singleton of x is $(L+x)+y$, which also reduces to $L+x+y$. So these two classes are identical. It follows that either class y is identical to individual x, or else class y is a member of itself, both of which are impossible. This completes the *reductio*. QED

Lassos correspond one-one to singletons, and therefore to members of singletons. Our old questions were: What, if anything, do we know about the nature of singletons? How is a singleton related to its member? Our new questions are: What, if anything, do we know about the nature of lassos? How is a lasso related to the thing it fits? We are no better off if we adopt the Lasso Hypothesis.

In fact, nothing changes at all. Whether we accept the Lasso Hypothesis or whether we reject it, we still say that to each x that has a singleton, there corresponds something y that is wholly distinct from x, and also there corresponds the fusion $y+x$. We could call y the 'lasso' and $y+x$ the 'singleton', and so accept the Lasso Hypothesis. Or we could call y the 'singleton' and thereby reject the Lasso Hypothesis, and then we could call $y+x$ the fusion (perhaps mixed) of the singleton and its member. What's in a name?

2.6 *Ramsifying out the Singleton Function*

We know nothing, so I lament, about the nature of the primitive relation between things and their singletons. What we do know, though, is that this relation satisfies certain structural conditions set forth in the axioms of set theory; and that it satisfies certain other conditions, for instance, that ordinary things such as cats, quarks, and space-time points should turn out to be among the things that have no singletons as parts. Might that be all we need to know? Paul Fitzgerald has suggested that it is.[9] (Or rather, he has suggested this for the membership relation generally. But if I am right that x is a member of a class y just when x is a member of a singleton that is part of y, and if I am right that the relation of part to whole is not itself mysterious, then we can confine our attention

---

[9] Paul Fitzgerald, 'Meaning in Science and Mathematics' in *PSA 1974*, ed. R. S. Cohen *et al.* (Reidel, 1976), section IV.

The Trouble with Classes

to the special case of member and singleton.) We needn't pretend to speak unequivocally of *the* function that takes members to singletons. Rather, any function that conforms to the appropriate conditions shall count as *a* singleton function. The content of set theory is that there exists some such function.

Axiomatize set theory with 'singleton' as the only set-theoretical primitive. Write down a system of structural, mathematical axioms chosen to yield standard iterative set theory. Make the axioms strong enough to rule out Skolemized interpretations with a countable domain, since these are obviously unintended. (We shall see in section 4.2 what such an axiom system should look like. Besides 'singleton' and elementary logic, there must of course be other primitive apparatus. But we may take that to be interpreted once and for all. It will be neither set-theoretical nor mysterious.) Add unofficial axioms as well, embodying those of our customary metaphysical opinions about classes that deserve credence. Include, for instance, an axiom stating that ordinary things – cats, etc. – are individuals, rather than classes or mixed fusions. Omit, however, those unofficial axioms which declare that singletons are outside space and time, that they are ethereal, or that they are platonic. Now conjoin all the axioms, mathematical and unofficial, into one single sentence:

...singleton...singleton...singleton...

The idea is to regard the word 'singleton' here not as an unequivocally meaningful primitive, but as a function-variable bound by an invisible quantifier. To give the content of the conjoined axioms more explicitly, we take the Ramsey sentence:

For some S; ...S...S...S...

Ramsifying out the Singleton Function

And if the word 'singleton' appears in some other sentence,

> Possum's singleton has a singleton

we should supply the quantifier and the structural conditions that restrict it, thus:

> For all S: if...S...S...S..., then Possum's S has an S.

(Or better, thus:

> For some S: ...S...S...S...; and
> for all S: if...S...S...S..., then Possum's S has an S

so that the existence of suitable values of S is affirmed rather than just presupposed. That brings our treatment of the conjoined axioms and our treatment of other sentences into line.)

Think of it this way: the word 'singleton' is highly equivocal, since all interpretations of it that satisfy the conjoined axioms are equally 'intended' interpretations. Though it appears in the guise of a primitive constant, it is no better than a variable – albeit a variable restricted to a certain limited range of admissible values. The equivocal sentence about Possum is true *simpliciter* iff it is true on all its intended interpretations (and there are some); equivalently, true on all admissible values (and there are some) of its disguised variable.[10]

---

[10] On tolerating equivocation, see van Fraassen, 'Singular Terms, Truth-Value Gaps and Free Logic' on supervaluations. On Ramsification, see F. P. Ramsey, 'Theories', in *Foundations* (Routledge and Kegan Paul, 1978); or Rudolf Carnap, *Philosophical Foundations of Physics* (Basic Books, 1966), section 26.

The Trouble with Classes

A parallel 'structuralist' account of arithmetic is familiar and attractive. We have the primitive notion 'successor'. (It can be our only arithmetical primitive. *Numbers* are those things that have successors. *Zero* is the number which is not a successor.) We have structural conditions, set forth in the Peano axioms, that the successor function is supposed to satisfy. But beyond that, alas, we haven't a clue about the nature of the succesor function. (Or maybe we are willing to reduce arithmetic to set theory, and claim – *pace* my present lamentations – to understand that. Then our predicament is that we have too many conflicting ideas about what the successor function might be, and no way to choose.) The structuralist's solution is to Ramsify out the successor function. He says that the word 'successor' is equivocal, no better than a (restricted) variable: any interpretation of it that satisfies the Peano axioms shall count as 'intended'. The content of the axioms is just that there exists some intended interpretation. A sentence in which the word 'successor' occurs is true *simpliciter* iff it is true on all intended interpretations; that is, true for all admissible values of its disguised variable.

The structuralist about arithmetic needn't scratch his head about the unknown nature of the number-successor relation; the structuralist about set theory needn't scratch his head about the unknown nature of the member-singleton relation. *Any* function that satisfies the stipulated conditions will do. We talk equivocally about all such functions. There's no secret fact about which one is the right one, the one we really mean. So says the structuralist.

I'm not content. Structuralism is not a such a bad position about arithmetic, though it would be better to respect our naive conviction that 'successor' has an unequivocal meaning. But structuralism for set theory does not fare so well.

In the first place, even if we could remove all mystery about the nature of the member-singleton relation, we would

Ramsifying out the Singleton Function

still have a mystery about the nature of the singletons. Note that we can say this, even if 'singleton' means nothing definite. For on all intended interpretations of 'singleton' alike, the same things will count as singletons: namely, all the mereological atoms, except for those that are parts of the ordinary things that, by axiomatic stipulation, were supposed to count as individuals.

The structuralist might retreat by dropping the unofficial axiom which says that singletons may not turn out to be among the atomic parts of ordinary things. Then *some* of the things that pass for singletons under some intended interpretations might be ordinary and unmysterious. But not nearly enough of them.

Here structuralism about arithmetic is better off. If we want enough ordinary things to make room for an intended interpretation of 'successor', countably many will do. It is easy to believe there are enough. Probably there are countably many space-time points (or point-sized momentary parts of particles) in this world alone, not to mention what the other worlds may have to offer. Maybe in this world, and surely in all the worlds together, there are at least countably many quarks, and countably many cats. So, unless we stipulate otherwise when we lay down the axioms, the things that pass for numbers under various intended interpretations of 'successor' might secretly be ordinary things, points or quarks or cats.

But standard iterative set theory – let alone large cardinal axioms! – makes far stronger demands on the size of Reality. There must be more than countably many atoms, more than continuum many, more than two to the continuum power many, more than two to the two to the continuum,...and all the demands in this sequence are but the barest beginning. It is hard to think where there could be that many ordinary atoms. So if there are enough things, and in particular enough atoms, to make room for an intended interpretation of 'singleton', then the overwhelming majority of these things must be *extra*ordinary. The preponderant part of Reality must consist of unfamiliar, unobserved things, whose existence would have gone unsuspected but for our acceptance of set theory, and of whose intrinsic nature we know nothing. It's like the astronomers' problem of the missing mass of the cosmos – but it's far, far worse.

The structuralist might retreat once more. He might formulate the axioms in an elementary way that would not, after all, rule out the undemanding Skolemized interpretations.[11] Then if he claimed that those interpretations were, after all, intended, he might conclude that ordinary things were abundant enough to meet his needs. I protest: those interpretations are *not* intended, and to say otherwise is to mutilate mathematics to suit our philosophy. It's subtle mutilation, but it's still unacceptable. And it's no excuse that there is a philosophical problem about how we grasp the difference between intended and Skolemized interpretations of set theory. We do grasp it. Any philosophy of thought and language that says we can't thereby stands refuted.

No retreat, then. Set-theoretical structuralism must acknowledge the mystery of the missing mass. At most it may hope to remove the mystery of the member-singleton function.

But in the second place, what can it mean to say – *not* yet presupposing set theory when we say it – that there exists a suitable function? We understand what makes a function suitable as an interpretation of 'singleton', but what's a

---

[11] If he did this by taking all instances of certain schemata as axioms, and if he did not abbreviate by means of substitutional 'quantification', then his conjuction of axioms would be an infinitely long sentence. I see no harm in that.

Ramsifying out the Singleton Function

function? A set theorist will say that a function is a special case of a relation, and a relation is a class of ordered pairs. Sometimes it's a set of pairs; but in the case of the singleton function, it's a proper class. But our structuralist cannot agree, not at this stage of the game, because he does not grant that the words 'class of ordered pairs' have any definite meaning.

Should the structuralist adopt some alternative theory of functions, and of relations generally? (1) There are various other set-theoretical analyses. But whatever advantages these may have, they are one and all unavailable at this stage of the game. (2) A theory which says that relations are *sui generis* individuals, but makes them class-like in all but name, can be expected to end in a predicament differing only in name from our present predicament. If relations are class-like, a plausible new 'Main Thesis' will say that the parts of a relation are all and only its subrelations; and we will be left scratching our heads over what it means for a minimal subrelation, instantiated just once, to relate one thing to another. Some neo-structuralist might propose to Ramsify out this primitive notion.... (3) A very different sort of theory of relations as *sui generis* individuals treats them as elements of being, building blocks of nature, constitutive of the qualitative character of things composed of interrelated parts. Spatiotemporal relations between the parts of a thing constitute that thing's shape and size, for instance. But we cannot explain the member-singleton relation in terms of any known aspects of the qualitative character of things, and it seems over-bold to respond by positing new, hidden aspects of qualitative character. What's good about structuralism is that it would allow us to get by (at the cost of losing uniqueness) with the sort of relations that pair things up arbitrarily and have nothing to do with qualitative character. (4) Some talk of relations may just be substitutional 'quantification' over relational predicates. But the only relevant predicates we

The Trouble with Classes

have are 'member of' and 'singleton of', and if we are structuralists we will regard these as unsuitable substituends because they lack definite meaning. (5) Or we may quantify over relational concepts; but that won't help unless we already have at least one definite concept suited to be the concept of membership in a singleton. I cannot survey all the theories that our structuralist might conceivably try, but it seems that we may dismiss the obvious candidates. Whatever their merits otherwise, they do not meet his present need.

He may solve part of his problem if he says not 'there is a class of pairs' but just 'there are some pairs'. He can insist, with Boolos, that plural quantification is primitively intelligible and need not and cannot be reduced to singular quantification over classes or anything else. I agree (see section 3.2).

But how, at this stage of the game, is the structuralist entitled to speak of ordered pairs? We usually define pairing set-theoretically, in various well-known ways. These definitions are not available to the structuralist. He denies that the primitive notion of set theory — 'member' or 'singleton' — has any definite meaning, so 'pair' defined in set-theoretical terms is no better off. If we need a set-theoretical definition of 'ordered pair', set-theoretical structuralism can succeed only if we understand set-theoretical notions to begin with. Only if we don't need it can we have it.

The structuralist might claim to have a primitive understanding of 'ordered pair'. But if so, why bother to be a structuralist? If he has pairing, he has self-pairing. And if he has self-pairing, he might as well define the singleton of x as the self-pair of x and x. Given reasonable axioms on primitive pairing, we would expect self-pairs to behave as singletons should. In requiring such axioms, he is no worse off than those of us who must lay down axioms directly on a

Ramsifying out the Singleton Function

primitive notion of singleton, or on a primitive notion of membership.

The game is not yet over. There might be some third way, neither set-theoretical nor primitive, to introduce pairing, and thereby to simulate quantification over relations, and thereby to get rid of the primitive notion of singleton. In fact this can be done, thanks to new work by John P. Burgess and A. P. Hazen. The appendix tells how. If we wish, we may have a new, structuralist set theory. First the general member-class relation gives way to the special case, member-singleton; then we Ramsify out member-singleton; and so we are left with no set-theoretical primitive at all. Farewell, then, to my complaints that the member-singleton relation is ill-understood. No worries, that which is not there requires no understanding.

Unfortunately, the structuralist redemption of set theory is not retroactive. Structuralist set theory really is new. It would be high-handed anachronism to claim that set theory was structuralist all along. It took the work of Burgess and Hazen to turn set-theoretical structuralism into an available option (if I am right to dismiss versions of structuralism that purport to quantify over relations taken as *sui generis* individuals). Before mid-1989, nobody even knew how to be a set-theoretical structuralist. To be a structuralist, you quantify over relations; to be a set-theoretical structuralist, you do so before you are entitled to the resources of set theory; and that is what we've only now learned how to do.

Arithmetical structuralism is a different case. Mathematicians have long known that the universe of mathematical objects affords countless functions that satisfy the structural conditions for a successor function. These functions are sets (set theory itself being taken for granted) so there is no problem about quantifying over them. The technical prerequisites

The Trouble with Classes

for Ramsifying out the successor function have long been common knowledge. So it is not anachronistic to say that arithmetic, as understood among mathematicians, is already structuralist. (Laymen are another question.) It is even somewhat plausible. Or at least it is somewhat plausible to say that arithmetic is not determinately not structuralist: even if mathematicians seldom or never avow arithmetical structuralism, at least it fits, well enough, the ways they mostly talk and think. Not so for set-theoretical structuralism. It does not fit present-day mathematical practice, because that practice does not include any knowledge of how to Ramsify out the singleton function.

To some extent, set-theoretical structuralism changes the subject. If we want to examine set theory as we find it now, we have to concede that it claims a primitive understanding of membership; and even when the general case of membership reduces via mereology to the special case of membership in singletons, still that special case is primitive. Despite all my misgivings over the notion of singleton, I am not fully convinced that structuralist revolution is the right response. I want to carry on examining set theory as we find it. Therefore I leave structuralism as unfinished business. Our working assumption for the rest of this book, until the appendix, shall be that the member-singleton relation is indeed primitive.

2.7 Metaphysics to the Rescue?

We can expect to hear from several systematic metaphysicians, each one offering to clarify the notion of singleton by subsuming it, in his own way, in his own system.

The first of them, of course, is the orthodox set theorist who tells us that a singleton is a special case of a class. True. Not helpful.

Metaphysics to the Rescue?

Next comes the one who tells us that pairing is primitive and that Possum's singleton is just the pair of Possum and Possum. False, I guess. Why favour the novel reduction of set theory to pairing (plus mereology) over the standard reduction of pairing to set theory?

Next comes one who tells us that Possum's singleton is Possum's *haecceity*: a special, non-qualitative property that can belong only to Possum.[12] True, I say, because it's part of a set-theoretical conception of properties – a property is any class of possibilia, and Possum's singleton is one such class – without which 'non-qualitative property' would be a contradiction in terms. But this does more to explain 'haecceity' then 'singleton'.

Next comes one who tells us that Possum's singleton is an abstraction from several coextensive qualitative properties, each one had by Possum alone. By 'property' he does not mean a class of actual and possible instances; by 'abstraction' he does not mean the taking of equivalence classes. My most urgent worry is that I have no guarantee that there is any qualitative property had by Possum alone. And if there is – if he is indeed a very special cat – it may be otherwise for other things that have singletons.

Next comes one who tells us that for each thing that has a property, there is a particular case of the property, a *trope*, that belongs to it alone. If Possum does have an exact duplicate, the two cats have duplicate tropes, not identical tropes. Then maybe Possum's singleton is one of his tropes: his own

---

[12] John Bigelow, *The Reality of Numbers: A Physicalist's Philosophy of Mathematics* (Oxford University Press, 1988), section 16, presents a theory of sets in general as plural haecceities. Thence we have the special case that a singleton is a singular haecceity. I do not know whether Bigelow could consistently say that a plural haecceity is a fusion of singular haecceities, and thereby endorse our Main Thesis. Anyhow, he doesn't.

The Trouble with Classes

particular self-identity, maybe, or perhaps his own particular cathood.[13] But even if we should believe in some tropes, particular self-identities seem *de trop*, and so do mereologically atomic tropes had by thoroughly non-atomic cats.

Last comes one who tells us that the world is the totality of facts, not of things; that although facts do in some sense have 'constituents', they are composed of these constituents in an altogether unmereological way, and are – at least sometimes – mereologically atomic; and that Possum's singleton is to be identified with some fact about him. Perhaps it should be the (mereologically atomic) fact that he is self-identical, or that he is a cat. Or perhaps it should just be the fact that he is one single thing, a unit; then unit classes would turn out to be facts of unithood.[14] Again, even granted an ontology of facts, I suspect it ought to be too sparse to afford atomic facts

---

[13] On tropes in general, see Williams, 'On the Elements of Being'; and Campbell, *Abstract Particulars*. Williams proposed the identification of singletons with self-identity tropes in lectures at Harvard *circa* 1963; but very tentatively, and as part of a broader plan that conformed to our First Thesis but not to our Second Thesis. The class of Magpie and Possum, for instance, would on Williams' plan be the fusion of three tropes: Magpie's self-identity (her singleton), Possum's self-identity (his singleton), and also the particular non-identity between Magpie and Possum.

[14] See Peter Forrest and D. M. Armstrong, 'The Nature of Numbers', *Philosophical Papers*, 16 (1987), pp. 165–86; and Armstrong, 'Classes are States of Affairs', forthcoming in *Mind*, 100 (1991). The latter paper gives exactly the proposal now under consideration; the former differs in one relatively minor way, as follows. Although it takes a class to be a fusion of some facts (or 'states of affairs') pertaining to its several members, it allows that the same thing might belong to different classes in virtue of different facts pertaining to it. If x is an F and also a G, Fx might be the fact whereby x belongs to the class of Fs, but Gx might be the fact whereby x belongs to the class of Gs. Then neither fact is once and for all the singleton that is a common part of both classes.

about non-atomic things. Further, mereology looks to be the *general* theory of composition, not the theory of one special kind of composition. Therefore I find unmereological 'composition' profoundly mysterious. After expelling it from set theory, I scarcely want to welcome it back via the anatomy of facts.

I am no enemy of systematic metaphysics. Nor do I imagine that the reservations I've expressed are at all conclusive against the several proposals. However, I want to pose a challenge to them all.

We were stumped by several questions about singletons. Where, if anywhere, are they? What is their intrinsic nature? Do they differ qualitatively from one another? Is the relation of member to singleton founded on the qualitative nature of the relata, or is it more like a distance relation, or is it a mixture, or something else altogether?

Let each of the metaphysicians tell us his answer to these questions, if indeed his theory of singletons yields answers. If it does, well and good. Let's see it done. But if it doesn't, then his subsumption of singletons has not dispelled their mystery. Rather, his subsumption shows that the mystery of singletons also bedevils the classes, or pairs, or haecceities, or abstractions from qualitative properties, or tropes, or facts. If he has shown that, he has not helped us and he has done himself no favour.

2.8 Credo

Singletons, and therefore all classes, are profoundly mysterious. Mysteries are an onerous burden. Should we therefore dump the burden by dumping the classes? If classes do not exist, we needn't puzzle over their mysterious nature. If we renounce classes, we are set free.

The Trouble with Classes

No; for set theory pervades modern mathematics. Some special branches and some special styles of mathematics can perhaps do without, but most of mathematics is into set theory up to its ears. If there are no classes, than there are no Dedekind cuts, there are no homeomorphisms, there are no complemented lattices, there are no probability distributions, ... For all these things are standardly defined as one or another sort of class. If there are no classes, then our mathematics textbooks are works of fiction, full of false 'theorems'. Renouncing classes means rejecting mathematics. That will not do. Mathematics is an established, going concern. Philosophy is as shaky as can be. To reject mathematics for philosophical reasons would be absurd. If we philosophers are sorely puzzled by the classes that constitute mathematical reality, that's our problem. We shouldn't expect mathematics to go away to make our life easier. Even if we reject mathematics gently – explaining how it can be a most useful fiction, 'good without being true'[15] – we still reject it, and that's still absurd. Even if we hold onto some mutilated fragments of mathematics that can be reconstructed without classes, if we reject the bulk of mathematics that's still absurd.[16]

---

[15] As in Hartry Field, *Science Without Numbers* (Princeton University Press, 1980).

[16] Does set-theoretical structuralism reject mathematics? It may seem not: it demands no change in the practice of mathematics, it counts the theorems as true, it says that classes exist, and it grants that speaking of singletons and their members makes sense. But it does demand wholesale reinterpretation. Whether acceptance-only-under-reinterpretation is tantamount to rejection is a vague matter – it depends on how drastic the reinterpretation is. I myself would not call the structuralist reinterpretation a rejection of mathematics; I do not dismiss it as absurd. Still, it does challenge the established understanding of mathematics for philosophical reasons, and all such challenges deserve suspicion.

Credo

That's not an argument, I know. Rather, I'm moved to laughter at the thought of how *presumptuous* it would be to reject mathematics for philosophical reasons. How would *you* like the job of telling the mathematicians that they must change their ways, and abjure countless errors, now that *philosophy* has discovered that there are no classes? Can you tell them, with a straight face, to follow philosophical argument wherever it may lead? If they challenge your credentials, will you boast of philosophy's other great discoveries: that motion is impossible, that a Being than which no greater can be conceived cannot be conceived not to exist, that it is unthinkable that anything exists outside the mind, that time is unreal, that no theory has ever been made at all probable by evidence (but on the other hand that an empirically ideal theory cannot possibly be false), that it is a wide-open scientific question whether anyone has ever believed anything, and so on, and on, *ad nauseam*?

Not me! And so I have to say, gritting my teeth, that somehow, I know not how, we do understand what it means to speak of singletons. And somehow we know that ordinary things have singletons, and singletons have singletons, and fusions of singletons sometimes have singletons. We know even that singletons comprise the predominant part of Reality.

3

A Framework for Set Theory

3.1   *Desiderata for a Framework*

Let us subdue our scepticism, and have faith in the teachings of set-theoretical mathematics. Let us accept the orthodox iterative conception of set, including the part of it that escapes elementary formulation. We want to end with nothing out of the ordinary.

But let us henceforth begin with singleton rather than membership as the primitive notion of set theory. We can leave it to mereology to make many-membered classes by fusing together singletons. And in formulating set theory, let us disentangle the part that characterizes the notion of singleton from the part that is just mereology.[1] (Let us separate Mr Hyde from Dr Jekyll.) Before we advance to set theory itself,

---

[1] Bunt, *Mass Terms and Model-Theoretic Semantics*, also formulates set theory with the part–whole relation and the member–singleton ('unicle') relation as primitive. But both primitives figure together in his axioms almost from the beginning; there is no attempt to set up mereology first and then add on the theory of singletons.

A Framework for Set Theory

we must have in place the framework to which we shall add it.

This framework will be topic-neutral, as logic is. It will be devoid of set theory, as logic is. It will be ontologically innocent, as logic is. It will be fully and precisely understood, as logic is. I would have liked to call it 'logic', in fact. But that is not its name, and, with names, possession is nine points of the law.

It will not be quite like logic, though. It will have a lot of mathematical power, as is shown in the appendix. (I commend it to those who dare to cut their mathematics to suit their philosophy.) In return, it will not admit of complete axiomatization. So we shall help ourselves to evident principles of the framework as needed, not stopping to make any official choice among its incomplete fragments. Our interest in axioms will concern not the framework, but rather the set-theoretical addition.

3.2 Plural Quantification

Besides the elementary logical apparatus of truth functions, identity, and ordinary singular quantification, our framework also shall be equipped with apparatus of plural quantification. We shall have plural pronouns (that is, variables) and plural quantifiers to bind them. We shall have a copula to link singulars with plurals. When we go beyond the vocabulary of the framework itself, we may have plural terms, and we may have predicates or functors (defined or primitive) that take plural arguments. This apparatus is not common in formal languages, but we know it well as masters of ordinary English. Let us remind ourselves of what we know how to say.[2]

---

[2] In this section I closely follow the lead of George Boolos, 'To Be is To Be the Value of A Variable (or To Be Some Values of Some Variables)'

Plural Quantification

A famous example: 'Some critics admire only one another.' Or in less abbreviated form: 'There are some critics such that each of them admires only other ones of them.' Or in long-winded regimentation: 'There are some things such that each of them is a critic; and such that for all x and y, if x is one of them and x admires y, then y is one of them and y is not identical to x.'

An example, from Boolos, to illustrate the power of plural quantification: Napoleon is Peter's ancestor iff, whenever there are some people such that each parent of Peter is one of them, and each parent of one of them is one of them, then Napoleon is one of them.

Examples to show the evident triviality of a principle of plural 'comprehension': If there is at least one cat, then there are some things that are all and only the cats. (Regimented: ...then there are some things such that, for all x, x is one of them iff x is a cat.) Likewise, if there is at least one set, then there are some things that are all and only the sets. If there is

---

and 'Nominalist Platonism'. I take it that I agree fully with him on substantive questions about plural quantification, though (as noted later) I would make less than he does of the connection with second-order logic. Other relevant writings on plurals include Bertrand Russell, *Principles of Mathematics* (Cambridge University Press, 1903), sections 70 and 74; Max Black, 'The Elusiveness of Sets', *Review of Metaphysics*, 24 (1971), pp. 614–36; Eric Stenius, 'Sets', *Synthese*, 27 (1974), pp. 161–88; Adam Morton, 'Complex Individuals and Multigrade Relations', *Noûs*, 9 (1975), pp. 309–18; Gerald J. Massey, 'Tom, Dick, and Harry, and All the King's Men', *American Philosophical Quarterly*, 13 (1976), pp. 89–107; D. M. Armstrong, *Universals and Scientific Realism*, vol. I, pp. 32–4; Richard Sharvy, 'A More General Theory of Definite Descriptions', *Philosophical Review*, 89 (1980), pp. 607–24; Peter M. Simons, 'Plural Reference and Set Theory', in *Parts and Moments: Studies in Logic and Formal Ontology*, ed. Barry Smith (Philosophia Verlag, 1982); Stephen Pollard, 'Plural Quantification and the Iterative Conception of Set', *Philosophy Research Archives*, 11 (1986), pp. 579–87.

at least one non-self-membered class, then there are some things that are all and only the non-self-membered classes.

Examples to illustrate the use of plural predicates: Isaac has written some books iff there are some things such that he has written them and they are books. He has written more than 450 books iff, whenever there are some books and they are at most 450 in number, then he has written some book that is not one of them. Likewise, he has written uncountably many books iff, whenever there are some books and they are countable in number, then he has written some book that is not one of them. (I presuppose, safely I think, that there exists at least one book.) The plural 'they are books' and 'Isaac has written them' reduce instantly to the singular: 'each one of them is a book', 'Isaac has written each one of them'. The plural 'they are at most 450' reduces less instantaneously: 'for some $x_1, \ldots,$ for some x_{450}, each of them either is x_1 or ... or is x_{450}'. As for 'they are countable', that might reduce with the aid of apparatus not yet introduced — set theory, for instance — or might remain primitive.

Examples of plural terms: 'The Dubliners', 'The Clancy Brothers and Tommy Makem', 'Andy Irvine and Paul Brady', 'the fans of the Chieftains'. The first is a plural proper name, the rest are eliminable plural definite descriptions. The fans of the Chieftains are many, for instance, iff there are some things, and they are many, and, for all x, x is one of them iff x is a fan of the Chieftains. The Clancy Brothers and Tommy Makem sing together iff there are some things, and they sing together, and, for all x, x is one of them iff either x is one of the Clancy Brothers or x is Tommy Makem.

Examples to deter us from confusing plural quantification with substitutional 'quantification' over predicates, or with singular quantification over sets or classes or properties: We can say without apparent contradiction — and indeed truly — that there are some things such that for no predicate are they

Plural Quantification

all and only the satisfiers of that predicate. There are some things such that for no set are they all and only its members; and some things such that for no proper class are they all and only its members; and some things such that for no property are they all only its instances.

Indeed, a nominalist well might say, falsely by my lights but without apparent contradiction, that although there are some people such that Peter's parents are among them and Napoleon isn't, and such that the parents of any one of them are among them, yet there is no set or class of these people, and no property that distinguishes them from other people. There are just people, says our nominalist, and other individuals, many of them; but no sets or classes or properties.

It is customary to take for granted that plural quantification must really be singular. Plurals, so it is said, are the means whereby ordinary language talks about classes. According to this dogma, he who says that there *are* the cats can only mean, never mind that he professes nominalism, that there *is* the class of cats. And if you say that there *are* the non-self-membered classes, you can only mean, never mind that you know better, that there *is* the class of non-self-membered classes.

If that's what you mean, what you say cannot be true: the supposed class is a member of itself iff it isn't, so there can be no such class. (Likewise there cannot be the set of all non-self-membered sets; and if there are class-like what-nots that are not exactly classes or sets, there cannot be the what-not of all non-self-membered what-nots.) To translate your seemingly true plural quantification into a contradictory singular quantification is to impute error – grave and hidden error.

We may well look askance at this imputation. Even the singularist may. He might dodge it by combining some imaginative interpretation of others' words with some imaginative ontology. When you say that there are the non-self-membered classes, the singularist says:

> By the dogma, you must mean that there is something that contains all and only the non-self-membered classes. But this something may not be what you ordinarily call a 'class'. I think, and I say that really you must think so too, that beyond the sets, and beyond the ordinary proper classes, there are some truly awesome 'classes' – if I may call them that – which contain even the proper classes as members. You must have meant that there is an awesome 'class' that contains all and only the non-self-membered sets and proper classes. *That's* no contradiction. *I* believe it – and you do too.

So you splutter, and eventually manage to say that you are not now, and never have been, a believer in awesome 'classes'. He says that you have just affirmed the existence of one, by plural quantification understood according to the singularist dogma.

When the nominalist embraces plural quantification and repudiates sets, he plays into the singularist's hands. The singularist says: 'Your plural quantification corresponds to my singular quantification over sets of individuals. How do you differ from me?' When Boolos embraces plural quantification and repudiates proper classes, he plays into the singularist's hands.[3] The singularist says: 'Your plural quantification corresponds to my singular quantification over sets and proper classes. How do you differ from me?' And when you embrace plural quantification and repudiate awesome 'classes', it's the same again. The singularist says: 'Your plural quantification corresponds to my singular quantification over sets,

---

[3] See Michael D. Resnik, 'Second-order Logic Still Wild', *Journal of Philosophy*, 85 (1988), pp. 75–87.

Plural Quantification

proper classes, and awesome "classes". How do you differ from me?'

The answer is the same every time. Maybe we're right in our various repudiations, maybe we're wrong. Suppose you repudiate sets. But there are still individuals, and they are available to be plurally quantified over. You may truly say that there are some things that are the cats. If you're right and there are no sets, then your true existential plural quantification over individuals cannot be the true existential plural quantification over sets that the singularist says it is – because *ex hypothesi* that singular quantification is false. If you're wrong and there are sets after all, proceed to the next case.

Suppose there are sets, but you still repudiate proper classes. If so, sets are available to be plurally quantified over. You may truly say that there are some things that are all the sets, or that there are some sets with no bound on their ranks, or that there are all the sets that are non-self-members. But there is no set of these sets. If you're right and there are no proper classes, then your true existential plural quantification over sets cannot be the true singular quantification over proper classes that the singularist says it is – because *ex hypothesi* that singular quantification is false. If you're wrong and there are proper classes after all, proceed to the next case.

Suppose there are proper classes, but you still repudiate awesome 'classes'. If so, proper classes are available to be plurally quantified over. You may truly say that there are some proper classes, but there is no set or proper class that has any of them as members; or that there are some things that are all the sets and proper classes that are non-self-members. If you're right and there are no awesome 'classes', then your true plural quantification over proper classes cannot be the true singular quantification over awesome 'classes' that the singularist says it is – because *ex hypothesi* that singular quantification is false. If you're wrong and there are awesome 'classes' after all, proceed to the next case.

Suppose there are awesome 'classes'; then they are available to be plurally quantified over; you can truly say that there are some things that are all and only the sets and proper classes and awesome 'classes' that are non-self-members. Maybe the singularist says that way out beyond even the awesome 'classes'. . . .

And so it goes. But let's cut a long story short. *Whatever* class-like things there may be altogether, holding none in reserve, it seems we can truly say that there are those of them that are non-self-members. Maybe the singularist replies that some mystical censor stops us from quantifying over absolutely everything without restriction. Lo, he violates his own stricture in the very act of proclaiming it!

We embrace plural quantification over *all* the things there are – even those, if any, that we may now be wrongly repudiating. How do we differ from the singularist, and from ourselves as he interprets us? We differ about what's true under any one fixed hypothesis about what there is and what there isn't.

The singularist dogma has the weight of common opinion in its favour. Quantification would be a simpler affair if it were true. Those are two weighty reasons to believe it. But not weighty enough. It imputes deviousness where there is no deviousness. It imputes ontological commitment where there is no ontological commitment. It imputes falsehood where there is no falsehood, and contradiction where there is no contradiction – unless instead it imputes restriction where there is no restriction, and ever-more-extravagant ontological commitment. Reject it, therefore. Plural quantification is irreducibly plural. It is not ordinary singular quantification over special plural things – not even when there are special

Plural Quantification

plural things, namely classes, to be had. Rather, it is a special way to quantify over whatever things there may be to quantify over. Plural quantification, like singular, carries ontological commitment only to whatever may be quantified over. It is devoid of set theory and it is ontologically innocent.

It may help to break the grip of singularism if we see how *some* plural quantification can be translated without dragging in any sets, classes, or properties. In the case of the prolific author, we quantified plurally over books. Let us eliminate the plurals and still quantify just over books. 'Whenever there are at most 450 books, some book by Isaac is not among them' becomes 'For any book x_1, for any book x_2, ..., for any book x_{450}, some book by Isaac is not x_1 and not x_2 and...and not x_{450}'. One plural quantifier over books gives way to 450 singular quantifiers over books. The next case is the same, but this time we need a language that allows infinite blocks of quantifiers and infinite conjunctions. 'Whenever there are countably many books, some book by Isaac is not among them' becomes 'For any book x_1, for any book x_2,..., some book by Isaac is not x_1 and not x_2 and...' where this time the ellipses abbreviate infinite sequences.

Can we say that plural quantifiers in general are abbreviations for long blocks of singular quantifiers, so long that their limited length never imposes any numerical restriction? I fear not. Before long, we'll have to concede that the abbreviated expressions exist only as set-theoretical sequences, and then I do not think we may mention them in explaining apparatus that purports to be devoid of set theory. What's worse, we'll eventually have to concede that the abbreviated expressions are too long to exist at all, even as sequences. And then they just don't exist, so we have no right to mention them at all. The reduction to quantifier-blocks is instructive, but yields only a skimpy fragment of plural quantification. The rest must remain primitive.[4]

Boolos identifies plural quantification with second-order quantification: monadic, full second-order quantification, with a restriction against vacuous variables. That will do to tell the experts what a formal language with plural quantification looks like, and what it can do. But I fear the identification may be misleading, though in no way has it misled Boolos himself. So I prefer to play it down.

(1) In so far as we already regarded second-order logic as set theory (or property theory) in sheep's clothing, it encourages singularism.

(2) It hints that the third, fourth, and higher orders cannot be far behind — but what might plurally plural quantification be? (Infinite blocks of plural quantifiers? — That will

---

[4] Morton, 'Complex Individuals and Multigrade Relations', treats plural quantifiers as blocks of singular quantifiers — safe enough, if he has set theory at his disposal and yet applies this treatment only to plural quantifiers over individuals, and if there are not too many individuals.

He joins this with the idea that we could analyse statements ostensibly about composite entities in terms of plural quantification over atoms alone. Then if we liked — Morton does not advocate it! — we could go on to deny that composition ever takes place. We could claim that 'the cat is on the mat' is true in virtue of the arrangement of many atoms, though strictly speaking there are no such composite entities as cats and mats. But even if we could save the truth of everyday statements, still this would be far from believable: 'I' are many atoms, not one composite thing. Further, it leaves us vulnerable to an argument akin to Malezieu's (see David Hume, *A Treatise of Human Nature* (1739), book I, part II, section II): if all composite entities are to be eliminated in favour of their parts, then we are forced to deny a priori that there can be such a thing as atomless gunk; because if there were, its parts in turn would be composite *ad infinitum*. Further, and fatally for purposes of my present project, it gives us no way to quantify plurally over mereological fusions.

be only a skimpy third order, and no start at all on the fourth.)

(3) It also hints that polyadic second-order quantification cannot be far behind. But, in fact, that takes additional resources. We need plural quantification over ordered pairs, triples, etc.; and we need some method of pairing. Even the methods of the appendix, which do without set theory (and primitive pairing), still require mereology and an infinite supply of atoms.

(4) Second-order quantification, unless especially restricted against vacuity, corresponds not to 'there are one or more things' but rather to 'there are zero or more things'. Because of this mismatch, regimented translations from unrestricted second-order logic into English come out abominably convoluted. Certainly there's something that's hard to understand! The hasty reader may think it's the plural quantification that's to blame.

3.3 Choice

Even before we reach mereology, let alone the set-theoretical notion of singleton, we have some of the power that's normally credited to set theory. We have already seen how to take ancestrals. Also, we can state – and should, of course, affirm – versions of the Axiom of Choice. Here are two. Both are schemata. To affirm them is to affirm whatever sentences can be made by filling in the blank, using any vocabulary that may later come to hand (adjusting the number of the verb), and then prefixing universal quantifiers, singular or plural, to bind any free variables.

> *First Choice Schema*: If there are some things, and each of them...some things, and no two of them...the same

A Framework for Set Theory

things, then there are some things such that each of the former things...exactly one of the latter things.

Second Choice Schema: If nothing ... itself, and if whenever $x ... y$ and $y...z$ then $x...z$, and if there are some things such that each of them ... another one of them, then also there are some of those things such that (1) among the latter things also, each one...another one; (2) whenever x and y are two of the latter things, then either $x...y$ or $y...x$.

The Second Choice Schema is a mouthful, but it says roughly this: if there is a partial ordering with no last term, there is a linear subordering of it still with no last term. But only roughly; because the Schema carries no commitment to any such entities as orderings.

3.4 Mereology

To finish our framework, we add the apparatus of mereology.[5] We have a choice of primitives. We could begin with 'part' and go on as follows.

---

[5] Mereology was developed by Stanislaw Leśniewski in several papers in Polish in 1916 and in 1927–1931: 'Podstawy Ogólnej Teoryi Mnogości I', *Prace Polskiego Koła Naukowego w Moskwie: Sekcya Matematyczno-przyrodnicza*, 2 (1916); and 'O Podstawach Matematyki'. *Przegląd Filozoficzny*, 30 (1927), pp. 164–206; 31 (1928), pp. 261–91; 32 (1929), pp. 60–101; 33 (1930), pp. 77–105; and 34 (1931), pp. 142–70. The 1927–1931 papers appear in abridged English translation as 'On the Foundations of Mathematics', *Topoi*, 2 (1983), pp. 3–52.

Leśniewski's mereology is founded upon, and includes, his Ontology. In Ontology there is a kind of quantification that can be interpreted, plausibly but without benefit of direct textual support, as plural; and

Mereology

Definitions: *x* and *y* *overlap* iff they have some common part. Iff not, they are *(entirely) distinct*.

(Those who hijack the word 'distinct' to mean merely 'non-identical' may say '*disjoint*' to mean 'non-overlapping'.)

Definition: Something is a *fusion* of some things iff it has all of them as parts and has no part that is distinct from each of them.

Or we could begin with 'overlaps', 'distinct', or 'fusion' instead.

---

there is a copula that can be read 'is one of'. (Here the plural subsumes not only the singular but also the empty: 'some things' must mean 'some zero or one or more things'.) See P. M. Simons, 'On Understanding Leśniewski', *History and Philosophy of Logic*, 3 (1983), pp. 165–91. Under this pluralist reading, Leśniewski's original version of mereology includes the whole of my 'framework'.

(Beware: Leśniewski thinks of mereology as affording an interpretation of the language of set theory. His term 'collective class' corresponds to our term 'fusion', not to the term 'class' in any sense now current.)

For other expositions of mereology in English, see A. N. Prior, *Formal Logic* (Oxford University Press, 1955), section III.4; Alfred Tarski, 'Foundations of the Geometry of Solids' in *Logic, Semantics, Metamathematics* (Oxford University Press, 1956); Tarski's Appendix E in J. H. Woodger, *The Axiomatic Method in Biology* (Cambridge University Press, 1937); Nelson Goodman and Henry Leonard, 'The Calculus of Individuals and Its Uses', *Journal of Symbolic Logic*, 5 (1940), pp. 45–55; Goodman, *The Structure of Appearance*, chapter II; Rolf A. Eberle, *Nominalistic Systems* (Reidel, 1970); and, for a very thorough survey, Peter Simons, *Parts: A Study in Ontology* (Oxford University Press, 1987), chapters 1 and 2.

If mereology is the general theory of part and whole, then it is a fragment of the 'ensemble theory' of Bunt, *Mass Terms and Model-Theoretic Semantics*. Bunt himself, however, reserves the term 'mereology' for a restricted theory of part and whole, applying only to individuals (see pp. 71–2, 288–301).

A Framework for Set Theory

Definitions: x is *part* of y iff everything that overlaps x also overlaps y; or iff everything distinct from y also is distinct from x; or iff y is a fusion of x and something z.

The basic axioms of mereology, besides whatever it may take to close the circle of interdefinition, are three.

Transitivity: If x is part of some part of y, then x is part of y.

Unrestricted Composition: Whenever there are some things, then there exists a fusion of those things.

Uniqueness of Composition: It never happens that the same things have two different fusions.

In view of the second and third axioms, we are entitled to say 'the fusion of', and rely on it to be a functor defined for any plural argument whatever.

These axioms do not settle all questions that can be raised in the language of mereology. Does Reality consist entirely of atomless gunk? Entirely of atoms? Or some of each? When we add set theory, then we shall conclude that there are atoms – namely, the singletons – but at this point we don't say. It will remain undecided whether there is gunk. And if there are atoms, how many are there? Using the resources of elementary logic, we can express such hypotheses as that there are more than seven, or that there are exactly seventeen, or that there are fewer than eighty. Using the resources of plural quantification, as we shall soon see, we can express hypotheses that discriminate between infinite sizes. But the basic axioms of mereology are silent about which of these hypotheses are true.

3.5 Strife over Mereology

I myself take mereology to be perfectly understood, unproblematic, and certain. This is a minority opinion. Many philosophers view mereology with the gravest suspicion.

Sometimes they suspect that originally the notion of part and whole was understood not as topic-neutral, but rather as a spatiotemporal – or merely spatial – notion. They conclude that any application of it to things not known to be in space and time is illicit. The original idea, supposedly, was that x is part of y iff y is wherever x is.

That is wrong thrice over. In the first place, it is common enough, and not especially philosophical, to apply mereology to things without thinking that they are in space and time. Trigonometry is said to be part of mathematics; God's foreknowledge is said to be part of His omniscience. In the second place, the proposed definition can go wrong even for things that are in space and time. Suppose two angels dance forever on the head of one pin. At every moment, each occupies the same place as the other. Still, they are two distinct proper parts of the total angelic content of their shared region. Suppose it turned out that the three quarks of a proton are exactly superimposed, each one just where the others are and just where the proton is. (And suppose the three quarks each last just as long as the proton.) Still the quarks are parts of the proton, but the proton is not part of the quarks and the quarks are not part of each other. Suppose a material thing occupies a region of substantival space-time; it does not follow (though it just might be true) that the region is part of the thing, still less that the region and the thing are parts of each other and therefore identical. In the third place, the proposed definition presupposes a prior notion of part and whole, because 'wherever *x* is' had better mean 'wherever some part of *x* is'.

Sometimes, philosphers' suspicions against mereology amount to guilt by association. Not just any controversial thesis that uses mereological notions and wins the adherence of some mereologists is a thesis of mereology *per se*. (1) Mereology is silent about whether all things are spatiotemporal. (2) It is silent about whether spatiotemporal things may have parts that occupy no less of a region than the whole does – for instance, about whether the quarks that make up the proton might be exactly superimposed; or about whether universals *in rebus*, wholly present wherever they are instantiated, might be among the proper parts of ordinary particulars; or about whether, alternatively, ordinary particulars are bundles of particular tropes, so that an electron would be composed of its charge-trope, its mass-trope, and a few more.[6] (3) Mereology is silent about whether something wholly present in one region may also be wholly present in another. For better or worse it does not forbid recurrent universals, or enduring things wholly present at different times, or a singleton atom that is where its extended member is by being at every point of an extended region, or the undivided omnipresence of God. (4) Finally, if something occupies a region, mereology *per se* does not demand that each part of the occupied region must be occupied by some part – proper or improper – of the occupying thing. If not, that's a second way for a singleton atom to be where its extended member is.

Sometimes temporal or modal problems about mereology

---

[6] On universals *in rebus*, see Armstrong, *Universals and Scientific Realism*; or the main system of Goodman, *The Structure of Appearance*. On tropes, see Williams, 'On the Elements of Being'; and Campbell, *Abstract Particulars*.

arouse suspicion. It seems that something and its proper part can be first different and later identical, or actually different but possibly identical. Then from the standpoint of the later time, or the unactualized possibility, it seems we have *one* thing that used to be, or could have been, not identical to itself!

I have discussed these problems elsewhere.[7] Briefly, they are instances of broader problems about temporary and accidental intrinsic properties. Those problems are not the wages of mereology. They arise for everyone. The temporal problems are best solved by accepting the doctrine of temporal parts. Other solutions (equally consistent with mereology) do exist, but have repugnant consequences, for instance, that I am a changeless soul, so not at all the sort of thing I think I am. The modal problems are best solved by denying the overlap of possible worlds. (Apart from shared universals, if such there be.) We can still say that our Hubert Humphrey – he himself! – wins the presidency at another world. But this can't mean that he himself is part of that world as well as this one. What can it mean? – Two theories explain it without overlap. My own theory says that Humphrey wins at another world by having a winning counterpart there, a counterpart not identical to Humphrey himself. A more popular rival theory says that possible worlds are abstract representations; some of them misrepresent our Humphrey as a winner, thereby ascribe to him the property of winning, and thereby confer upon him the property of winning at (better, winning according to) that other world. When a possible world misrepresents Humphrey as winning, no

---

[7] Lewis, *On the Plurality of Worlds*, chapter 4. For related discussions see Lewis, 'Counterparts of Persons and Their Bodies', *Journal of Philosophy*, 68 (1971), pp. 203–11; and 'Rearrangement of Particles: Reply to Lowe', *Analysis*, 48 (1988), pp. 65–72.

A Framework for Set Theory

Humphrey-as-misrepresented exists as part of that possible world, or anywhere else. What's true is that he exists *according* to that possible world.

Sometimes it is the axiom of Uniqueness of Composition that arouses suspicion. It says there is no difference without a difference-maker: x and y are identical unless there is something to make the difference between them by being part of one but not the other. Against this it might be said, for instance, that the two words 'master' and 'stream' are made of the same six letters. No; various things may happen, but none of them is a violation of the axiom. (1) There is a 'master'-inscription and there is a 'stream'-inscription, and each one is made of six letter-inscriptions. Then the 'a' in 'master' and the 'a' in 'stream' are not the same, though they are alike in shape. (2) The two words might be written crossword-fashion so that they share the same 'a'-inscription; but then the 'm'-inscriptions of the two words would not be the same. (3) One single inscription, made of six letter-inscriptions, might be both a 'stream'-inscription and also a 'master'-in-peculiar-order-inscription. (4) The letters might be movable pieces of plastic, formed first into a 'master'-inscription and then rearranged into a 'stream'-inscription. Then the two word-inscriptions are made of different temporal parts of the same six persisting letter-inscriptions. (5) Suppose, as I think wrongly, that the six plastic letters do not have temporal parts. Instead they persist by enduring identically, wholly present at different times. Then the unique thing composed of the six letters is a 'master'-inscription relative to one time and a 'stream'-incription relative to another. There's no kind of inscription that it is *simpliciter*; nor, on this supposition, do we have two different inscriptions of the two words. (6) Suppose the word-types 'master' and 'stream' are taken to be each the fusion of all its inscriptions; and likewise for letter-types. (To avoid confusion, we'd better take only those

Strife over Mereology

inscriptions which contrast with their surroundings.) Then the two word-types are each made of parts of the same six letter-types; but not the same parts. (7) Suppose instead that the word-types are taken as set-theoretic constructions out of the six letter-types – sequences, perhaps. Then indeed two things are generated set-theoretically out of the same six members. It is just this power to make the most of limited material that makes set-theoretic constructions worthwhile. But they are not cases of composition, so I claim, because what is going on is not just composition. Partly it is composition; partly it is the making of singletons, which makes one thing out of *one* thing, not out of many. I conclude that words and their letters, and parallel examples, are no threat to Uniqueness of Composition.[8]

Most of all, it is the axiom of Unrestricted Composition that arouses suspicion. I say that whenever there are some things, they have a fusion. *Whenever!* It doesn't matter how many or disparate or scattered or unrelated they are. It doesn't matter whether they are all and only the satisfiers of some description. It doesn't matter whether there is any set, or even any class, of them. (Here's where plural quantification

---

[8] But what of another way of taking the example? (8) A word-type is a structural universal instantiated by the inscriptions; it is made of simpler structural universals, namely the six letter-types together with relations of juxtaposition; these relations, as well as the letter-types, are involved alike in both word-types. This really would go against the principle – no misunderstanding this time. But to accept this as a counterexample, we must first accept a theory of structural universals which I take to be far from established. Rather than taking the case as an objection to the principle of Uniqueness of Composition, I go the other way: I object to structural universals exactly because they require some sort of unmereological 'composition' that violates the principle. See David Lewis, 'Against Structural Universals', *Australasian Journal of Philosophy*, 64 (1986), pp. 25–46.

pays its way, for better or worse.) There is still a fusion. So I am committed to all manner of unheard-of things: trout-turkeys, fusions of individuals and classes, all the world's styrofoam, and many, many more. We are not accustomed to speak or think about such things. How is it done? Do we really have to?

It is done with the greatest of ease. It is no problem to describe an unheard-of fusion. It is nothing over and above its parts, so to describe it you need only describe the parts. Describe the character of the parts, describe their interrelation, and you have *ipso facto* described the fusion. The trout-turkey in no way defies description. It is neither fish nor fowl, but it is nothing else: it is part fish and part fowl. It is neither here nor there, so where is it? – Partly here, partly there. That much we can say, and that's enough. Its character is exhausted by the character and relations of its parts.

I never said, of course, that a trout-turkey is no different from an ordinary, much-heard-of thing. It is inhomogeneous, disconnected, and not in contrast with its surroundings. (Not along some of its borders.) It is not cohesive, not causally integrated, not a causal unit in its impact on the rest of the world. It is not carved at the joints. But none of that has any bearing on whether it exists.

If you wish to ignore it, of course you may. Only if you speak with your quantifiers wide open must you affirm the trout-turkey's existence. If, like most of us all the time and all of us most of the time, you quantify subject to restrictions, then you can leave it out. You can declare that there just does not exist any such thing – *except*, of course, among the things you're ignoring.

Doing away with queer fusions by restricting composition cannot succeed, unless we do away with too much else besides. For many respects of queerness are matters of degree. But existence cannot be a matter of degree. If you say there is something that exists to a diminished degree, once you've said 'there is' your game is up. Existence is not some special distinction that befalls some of the things there are. Existence just *means* being one of the things there are, nothing else. The fuzzy line between less queer and more queer fusions cannot possibly coincide with the sharp edge where existence gives out and nothing lies beyond. A restriction on your quantifiers, on the other hand, may be as fuzzy as you please.[9]

3.6 Composition as Identity

So I claim that mereology is legitimate, unproblematic, fully and precisely understood. All suspicions against it are mistaken. But I claim more still. Mereology is ontologically innocent.

To be sure, if we accept mereology, we are committed to the existence of all manner of mereological fusions. But given a prior commitment to cats, say, a commitment to cat-fusions is not a *further* commitment. The fusion is nothing over and above the cats that compose it. It just *is* them. They just *are* it. Take them together or take them separately, the cats are the same portion of Reality either way. Commit yourself to their existence all together or one at a time, it's the same commitment either way. If you draw up an inventory of Reality according to your scheme of things, it would be double counting to list the cats and then also list their fusion. In general, if you are already committed to some things, you incur no further commitment when you affirm

---

[9] For a fuller statement of this argument against restricting composition, see Lewis, *On the Plurality of Worlds*, pp. 212–13. For a rejoinder, see Peter van Inwagen, 'When are Objects Parts?', *Philosophical Perspectives*, 1 (1987), pp. 21–47, especially pp. 40–5.

A Framework for Set Theory

the existence of their fusion. The new commitment is redundant, given the old one.

For the most part, if you are committed to the existence of a certain thing or things, and then you become committed to the existence of something that bears a certain relation to it or them, that is indeed a further commitment. If you incur a commitment to the lover of, or the next-door neighbour of, or the weight in grams of, or the shadow of, or the singleton of, something you were committed to already, you have made a further commitment. It is not redundant. But the relation of identity is different. If you are already committed to the existence of cat Possum, and then affirm that there exists something identical to Possum, that is not a further commitment. I say that composition – the relation of part to whole, or, better, the many-one relation of many parts to their fusion – is like identity. The 'are' of composition is, so to speak, the plural form of the 'is' of identity. Call this the Thesis of *Composition as Identity*. It is in virtue of this thesis that mereology is ontologically innocent: it commits us only to things that are identical, so to speak, to what we were committed to before.

In endorsing Composition as Identity, I am following the lead of D. M. Armstrong and Donald Baxter. Armstrong takes strict identity and strict difference as the endpoints of a spectrum of cases, with cases of more or less extensive overlap in between. Overlap subsumes part-whole as a special case: it may be x itself, or y itself, that is a common part of x and y. Two adjoining terrace houses that share a common wall

> are not identical, but they are not completely distinct from each other either. They are partially identical, and this partial identity takes the form of having a common part. Australia and New South Wales are not identical, but they are not completely

Composition as Identity

distinct from each other. They are partially identical, and this partial identity takes the form of the whole-part 'relation'.... Partial identity admits of at least rough-and-ready degree. Begin with New South Wales and then take larger and larger portions of Australia. One is approaching closer and closer to complete identity with Australia.[10]

Baxter puts it this way:

> The whole is the many parts counted as one thing. On this view there is no one thing distinct from each of the parts which is the whole. Rather, the whole is simply the many parts with their distinctness from each other not mattering. This...is not to deny the existence of the whole. It is merely to deny the additional existence of the whole....
>
> Suppose a man owned some land which he divides into six parcels. Overcome with enthusiasm for [the denial of Composition as Identity] he might try to perpetrate the following scam. He sells off the six parcels while retaining ownership of the whole. That way he gets some cash while hanging on to his land. Suppose the six buyers of the parcels argue that they jointly own the whole and the original owner now owns nothing. Their argument seems right. But it suggests that the whole was not a seventh thing.[11]

Indeed! I grant that no one of the six parcels *by itself* is identical to the original block of land. Still, there is a good sense in which the six parcels and the original block are the very same thing. They are it, it is them. You can't sell *them* without selling *it*, because you can't sell *it* without selling *it*.

---

[10] Armstrong, *Universals and Scientific Realism*, vol. II, pp. 37–8.
[11] Donald Baxter 'Identity in the Loose and Popular Sense', *Mind*, 97 (1988), p. 579. Also see Donald Baxter, 'Many-One Identity', *Philosophical Papers*, 17 (1938), pp. 193–216.

A Framework for Set Theory

A doubter might seek to trivialize Composition as Identity, thus:

> Of course the terrace houses are partially identical, and so are New South Wales and Australia, if by that you just mean that something that is part of one is identical to something that is part of the other. Of course the six parcels are identical to the original block, if by that you just mean that the fusion of the parcels is identical to the block. But that shows nothing about the innocence of mereology. You could do the same trick with any old relations. Possum is 'maternally identical' to Magpie, if by that you just mean that something that is mother of one is identical to something that is mother of the other. The six numbers 2, 8, 4, 7, 6 and 3 are 'on average' identical to the one number 5, if by that you just mean that the average of the six is identical to the one. So what?

But the doubter misses the point. Never mind that mereological relations can be equivalently restated so as to drag in identity. Anything can be, as he said. The real point is that the mereological relations (however restated) are something special. They are unlike the same-mother relation or the average-of relation. Rather, they are strikingly analogous to ordinary identity, the one-one relation that each thing bears to itself and to nothing else. So striking is this analogy that it is appropriate to mark it by speaking of mereological relations – the many-one relation of composition, the one-one relations of part to whole and of overlap – as kinds of identity.

[12] Here I part company with Baxter. Composition as Identity comes in different versions. My version, as just stated, is analogical. A stronger, and stranger, version would disdain mere analogy. It would insist that there is just one kind of identity, the ordinary one-one kind, and that composition involves this kind of identity. It is this view, or something very like it, that Baxter defends. His ingenious defence makes better sense of the view than I'd have thought possible; but in the end I'm unconvinced, and so rest content with mere analogy.

Composition as Identity

Ordinary identity is the special, limiting case of identity in the broadened sense.[12]

The analogy has many aspects. Four of them have come to our attention already. The first is the ontological innocence of mereology: just as it is redundant to say that Possum exists and something identical to him exists as well, so likewise it is redundant to say that Possum and Magpie both exist and their fusion exists as well.

Unrestricted composition is a second aspect of the analogy. If Possum exists, then automatically something identical to Possum exists; likewise if Possum and Magpie exist then automatically their fusion exists. Just as Possum needn't satisfy any special conditions in order to have something identical to him, so Possum and Magpie needn't satisfy any special conditions in order to have a fusion.

Uniqueness of Composition is a third aspect of the analogy. Just as there cannot be two different things both identical to Possum, likewise there cannot be two different fusions of Magpie and Possum. A kind of transitivity applies, with Magpie and Possum together as a plural middle term. If x is them, and they are y, then x is y.

The ease of describing fusions is a fourth aspect of the analogy. Describe Possum fully, and thereby you fully describe whatever is identical to Possum. Describe Magpie and Possum fully – the character of each, and also their interrelation – and thereby you fully describe their fusion. Likewise for relational description: specify the location of Magpie and Possum, and thereby you specify the location of their fusion. Specify the present ownership of the six parcels and thereby you specify the ownership of the original block of land.

A fifth aspect of the analogy has to do with multiple location. If Mary's lamb goes everywhere that Mary goes, and if this is so not just as a matter of fact but as a matter of absolute necessity, we have a highly mysterious necessary connection between distinct existences. But if it turns out that Mary and the lamb are identical, then there is no mystery at all about their inseparability. Likewise if it turns out that the lamb is part of Mary, and if Mary is *wholly* present wherever she goes, then again the inseparability is automatic, and in no way mysterious. In general, if a fusion is multiply located – wholly present at different places or times or possible worlds – then also its parts must be multiply located. Since it just is them, it cannot appear anywhere without them.

It would be nice to illustrate this by example. Perhaps I can, if there are multiply located universals. If a conjunctive universal P & Q is the fusion of its conjuncts P and Q,[13] then it is automatic and unmysterious that both P and Q must appear wherever P & Q does.

As for alleged examples that involve multiply located particulars, instead of universals, I definitely disbelieve in those. To be sure, the same road may have a lane on the hilly stretch that it lacks on the plain, the same man may have more teeth when young than when old, or he may actually have more toes than he might have had. But to these cases I apply the principle contrapositively. Because the thing varies with respect to its parts, I conclude that we do not have genuine multiple location. It is not the same thing wholly

---

[13] Armstrong, *Universals and Scientific Realism*, vol. II, p. 36, says that P is part of P & Q. There is a problem: the fusion of P and Q exists whether or not P and Q are coinstantiated, but according to Armstrong we have no conjunctive universal P & Q unless P and Q are coinstantiated. Does this show that conjunctive universals cannot be the fusions of their conjuncts? No; it only shows that if P and Q are not coinstantiated, their fusion is not *in that case* a universal. But we can perfectly well say that whether the fusion is a universal or not depends on whether P and Q are coinstantiated.

present at multiple locations. It is the same thing partially present, or different things wholly present.

This completes the analogy that I take to give the meaning of Composition as Identity. But alas, it has its limits. In the first place, I know of no way to generalize the definition of ordinary one-one identity in terms of plural quantification. We know that x and y are identical iff, whenever there are some things, x is one of them iff y is one of them. But if y is the fusion of the xs, then there are some things such that each of the xs is one of them and y is not; and there are some things such that y is one of them but none of the xs is. And in the second place, even though the many and the one are the same portion of Reality, and the character of that portion is given once and for all whether we take it as many or take it as one, still we do not really have a generalized principle of indiscernibility of identicals. It does matter how you slice it — not to the character of what's described, of course, but to the form of the description. What's true of the many is not exactly what's true of the one. After all they are many while it is one. The number of the many is six, as it might be, whereas the number of the fusion is one. And the singletons of the many parts are wholly distinct from the singleton of the one fusion. That is how we can have set theory.

Plural quantification is innocent: we have many things, we speak of them as many, in no way do we mention one thing that is the many taken together. Mereology is innocent in a different way: we have many things, we do mention one thing that is the many taken together, but this one thing is nothing different from the many. Set theory is not innocent. Its trouble has nothing to do with gathering many into one. Instead, its trouble is that when we have one thing, then somehow we have another wholly distinct thing, the singleton. And another, and another, . . . *ad infinitum*. But that's the price for mathematical power. Pay it.

A Framework for Set Theory

3.7 Distinctions of Size

Our framework, as it turns out, affords the means to define distinctions in the size of things. Up to a point, of course, this can be done in elementary logic: using identity, we can define 'thing with more than seventeen atoms' and the like. But with the resources of plural quantification and mereology, we can do much more.

We begin with the distinction between finite and infinite things. An infinite thing is one with infinitely many parts; in that case, its parts will admit of an endless ordering; a partial ordering will do; and we have one ready to hand, namely the relation of part to whole.

> *Definitions*: x is *infinite* iff x is the fusion of some things, each of which is a proper part of another. Otherwise x is *finite*.

Note that any bit of atomless gunk is infinite: it is the fusion of all its proper parts, each of which is a proper part of another. Likewise anything that has an atomless part is infinite. A thing that consists entirely of atoms, on the other hand, may be either finite or infinite. An obvious consequence of the definition is that an atom is finite.

There are other ways of defining infinity. One is Dedekind's: when we have infinitely many things, then all those things correspond one-one with only some of those things. This may seem useless until we have ordered pairs. Useless as a definition, yes; but at least we may affirm it as a principle of the framework.

> *Dedekind Schema*: If x is a proper part of y, and if each atom of y . . . exactly one atom of x, and if each atom of

Distinctions of Size

x is such that exactly one atom of y...it, then y is infinite.

(Note that if x includes all the atoms of y, leaving out only some atomless gunk, then we do not know that y has infinitely many atoms. But then y is infinite because of the gunk.)

Next, a relative distinction. We may take all the atoms of Reality, however many of them there may be, to set a standard of size. 'Size' here is measured by atoms; atomless gunk, if there is any, won't count. Roughly, I want to say that something is 'large' iff it has as many atoms as there are in all of Reality; otherwise 'small'. Without benefit of ordered pairs, I cannot say quite that. But I *can* say that something has at least as many atoms as there are in all the *rest* of Reality; which comes to the same thing, except in the special case – contrary to set theory, and to any physics that posits points of the space-time continuum – that Reality has only finitely many atoms.[14]

> *Definitions*: x is *large* iff there are some things such that (1) no two of them overlap, (2) their fusion is the whole of Reality, and (3) each of them contains exactly one atom that is part of x and at most one other atom. Otherwise x is *small*.

It follows immediately that any part of a small thing is small.

We may affirm principles of the framework that connect

---

[14] A price we pay for not saying just what we want to say concerns the smallness of atoms. We'd want it to come out that a single atom is small, unless it is the only atom; in other words, that an atom is small iff there are two or more atoms. But instead it turns out that an atom is small iff there are *three* or more atoms.

the distinction of infinite versus finite to the distinction of large versus small. If there is something infinite that consists entirely of atoms, then (1) any finite thing is small, and indeed (2) any fusion of a small thing with a finite thing is small. (The small thing may itself be finite, and if so the fusion is finite too; or the small thing may be small and infinite, and if so the fusion has the same small infinite size.)

We may also affirm, as a principle of the framework, this

> *Replacement Schema (singular version)*: If each atom of a thing x ... exactly one atom of a thing y, and if for each atom of y there is an atom of x that ... it, and if x is small, then y is small.

(That is, again, we affirm each sentence made by filling in the blank and prefixing universal quantifiers as required.) Set theorists beware: the schema is not set theoretical, and it in no way constrains the size of Reality.

I would like to introduce plural predicates 'few' and 'many', corresponding to the singular predicates 'small' and 'large', where again the atoms in all of Reality set the standard.[15] But without ordered pairs, I cannot do this in full generality; so my definition applies only to things whose fusion does not fill up too much of Reality. (Two halves of Reality, though only two, won't count as few!)

---

[15] Beware; the plural distinction of few and many differs in one way from the singular distinction of small and large. Unless there are but finitely many atoms, 'large' is a single size: all large things are equal, each having as many atoms as there are in all of Reality. But 'many' is not a single size. There are many atoms, and many fusions of atoms, but there are more of the latter than of the former. The atoms are barely many, the fusions of atoms are more than barely many.

Distinctions of Size

Definition: Suppose we have some things such that some large thing does not overlap any of them. Then they are *few* iff there is some small thing x, and there are some things, such that (1) x does not overlap the fusion of the former things, (2) each of the latter things is the fusion of one of the former things and one atom of x, (3) for each of the former things, one of the latter things is the fusion of it and one atom of x, and (4) no atom of x is part of two or more of the latter things. Otherwise they are *many*.

We connect our plural and singular predicates by affirming, as a principle of the framework, that something is small iff its atoms are few. And we may affirm, as a principle of the framework, this

Replacement Schema (plural version): Given some things, and given some other things (not necessarily different), if each of the former things . . . exactly one of the latter things, and if for each of the latter things there is one of the former things that . . . it, and if the former things are few, then the latter things are few.

Much more could be done, within the vocabulary of the framework, to define distinctions of size (see the appendix). Further, we could look for an elegant systematic theory that would yield, as axiom or as theorem, each of the 'principles of the framework' that I have affirmed in passing, and others besides. But we have done enough to be going on with. We have in store just what we shall need later.

4

Set Theory for Mereologists

4.1 The Size of Reality

Now that we have our framework, it is time for an end to our ontolgical innocence. We shall take on our commitments in two stages. First, holding the primitive notion of singleton still in abeyance, we shall assert that there is a very great deal of *something*. We can't say yet just what this great deal of somthing consists of, but of course – to jump the gun – we know it must consist predominantly of singletons.

We may affirm three hypotheses about the size of Reality. These are not innocent principles of the framework, but rather are forced upon us by our acceptance of orthodox, set-theoretical mathematics. The first two correspond, as we shall see later, to the set-theoretical axioms of Power Sets and Unions.

> *Hypothesis P*: If something is small, then its parts are few.

> *Hypothesis U*: If some things are small and few, their fusion is small.

Set Theory for Mereologists

Hypotheses P and U would be true if Reality were barely infinite, with a countable infinity of atoms. 'Small' and 'few' would then mean 'finite'. Hypothesis P would say, truly, that a finite thing has only finitely many parts. Hypothesis U would say, truly, that the fusion of finitely many finite things is still finite.

Each hypothesis also can be true at certain larger infinite sizes. Hypothesis P would be true, for instance, if the size of Reality were like the limit cardinal beth-omega. Hypothesis U would be true, for instance, if the size of Reality were like beth-one (the continuum), or aleph-one, or beth-seventeen. The problem comes in demanding that the two be true together. That happens, we saw, if Reality is countably infinite. It does not happen again until we reach an 'inaccessible' infinite size[1] that transcends our commonplace alephs and beths in much the same way that those transcend mere finitude. The story is well know as told in set theory; but our framework of plural quantification and mereology has enough power that we can tell it there too.

If we want orthodox mathematics, we cannot be content with mere countable infinity. We must demand a little more. But with Hypotheses P and U in place, to demand a little is willy-nilly to demand a lot.

Hypothesis I: Some fusion of atoms is infinite and yet small.

By itself, Hypothesis I is undemanding. It excludes only the smallest of infinite sizes. It is only when combined with Hypotheses P and U that it demands the high jump. What might there be uncountably and inaccessibly many of? Not

---

[1] 'Strongly' inaccessible.

Mereologized Arithmetic

cats, not quarks, not space-time points, it's safe to say! So now it's time for the primitive notion of singleton.

4.2 Mereologized Arithmetic

Our set-theoretical primitive can at first be taken as a two-place predicate: x is a singleton of y. But the first of our axioms for it,

> *Functionality:* Nothing has two different singletons.

straightway entitles us to treat it henceforth as a functor, and speak of *the* singleton of something. (A singleton, of course, is something that is the singleton of something.) Our next axiom specifies the domain of the singleton function.

> *Domain:* Any part of the null set has a singleton; any singleton has a singleton; any small fusion of singletons has a singleton; and nothing else has a singleton.

Our next axiom specifies that different things have distinct singletons, and that singletons are distinct from parts of the null set. The axiom demands distinctness in the strong sense of non-overlap.

> *Distinctness:* No two things have overlapping singletons, nor does any part of the null set overlap any singleton.

Distinctness in the sense of mere non-identity follows: nothing is the singleton of two different things, nor is any singleton part of the null set. Finally we have an axiom of induction: it says that we can get to anything there is by starting with the null set and its parts, and iterating the operations of singleton and fusion.

> *Induction*: If there are some things, if every part of the null set is one of them, if every singleton of one of them is one of them, and if every fusion of some of them is one of them, then everything is one of them.

In the axioms as stated, 'null set' looks like an unacknowledged second primitive. Let us set that right. By Induction, we have that everything is either a part of the null set, a fusion of singletons, or a fusion of a part of the null set and some singletons. Distinctness implies that the null set has no singletons as parts. Hence it follows from the axioms that the null set may be defined as the fusion of all things that have no singletons as parts, and that there is such a thing. So the axioms could have been rewritten just in terms of 'singleton'.

A singleton must be an atom.

Proof What could be a part of it? Only a part of the null set, a fusion of singletons, or a fusion of a part of the null set and some singletons — for those are the only things there are. But by Distinctness, it cannot have as a part either any part of the null set or any singleton other than itself. So it can have no part except itself.   QED

Once we know that singletons are atoms, we can show that something that consists entirely of atoms, namely the fusion of all singletons, is infinite.

Proof By Domain, there are singletons, for instance the singleton of the null set, and all singletons have singletons. But by Distinctness, some singletons, for instance the singleton of the null set, are not singletons of singletons. So if x is

Mereologized Arithmetic

the fusion of all singletons of singletons, and y is the fusion of all singletons, then x is a proper part of y. Further, each singleton has as singleton exactly one singleton of a singleton, by Functionality; that is, each atom of y has as singleton exactly one atom of x. Further, each singleton of a singleton is the singleton of exactly one singleton, by Distinctness; that is, each atom of x is such that exactly one atom of y has it as singleton. So y is infinite, by the Dedekind Schema.   QED

Once we know that something that consists of atoms is infinite, we know that anything finite is small. Singletons, being atoms, are finite. So singletons are small. Hence singletons themselves are subsumed under small fusions of singletons, and Domain looks to have a redundant clause.[2] The things that have singletons are exactly the parts of the null set and the small fusions of singletons.

Recall our previous definitions in terms of 'singleton', and let them be taken into mereologized arithmetic. We defined a *class* as a fusion of singletons, and the *members* of a class as things whose singletons are parts of that class. We defined an *individual* as anything that has no singletons as parts. We

---

[2] So why did I write it that way? Because I had to assume that each singleton had a singleton *before* I could establish that something consisting of atoms was infinite and hence that singletons were small. Without the assumption, I could have shown that there were *two* atoms: namely the null set and its singleton, if the null set is an atom, and otherwise the singletons of the null set and one of its proper parts. But I could not have shown that there were *three*; and only if there are three does it follow that atoms are small (see section 3.7, note 14).

To see this, note that if the clause 'any singleton has a singleton' were stricken from Domain, then all the axioms of mereologized arithmetic would be satisfied if there existed just two atoms, the null set and its singleton; their fusion; and nothing else. There would be no small fusions of singletons, hence no singletons thereof, because the solitary singleton, accompanied by just one other atom, would not be small.

defined the *null set* as the fusion of all things that have no singletons as parts – the same definition that now follows from our axioms. We defined an *urelement* as something that has no singletons as parts, but is part of something else of which the same is true, i.e. an individual other than the null set. We defined a *set* as either the null set or a class that has a singleton.

Accordingly, a class is a set iff it is small. A large class is a proper class, it lacks a singleton, and it cannot be a member of anything. The things that can be members are exactly the sets and the individuals; in other words, the small classes, the null set, and the urelements. Here is the principle of Limitation of Size, in an uncommonly explicit form. It is a strong form of the principle, following von Neumann: not only does it say that the proper classes are larger than all others, also it says that all proper classes are the same size – namely, the largest size.

4.3 Four Theses Regained

At the start, I advanced four intuitive theses about the mereology of classes. Arguing from these theses together with orthodox set theory, I concluded that the class of cats was the fusion of singletons of cats, that Possum was a member of just those classes that had his singleton as a part, and so forth. Now I am arguing in the opposite direction: starting from the axioms and definitions of mereologized arithmetic, plus principles of the framework, I want to bring us back to where we started. First we regain the original four theses. Then we shall regain orthodox set theory.

> *First Thesis*: One class is part of another iff the first is a subclass of the second.

Four Theses Regained

Proof Left to right. Suppose class x is part of class y and z is a member of x. By our present definition of membership, z's singleton is part of x and hence part of y. So z is member of y. Right to left. Prove the contrapositive. Suppose class x is not a part of class y. Since x is a fusion of singletons, by definition of class, and singletons are atoms, some singleton is part of x but not of y. Let it be z's singleton; by definition of membership, z is a member of x but not of y, so x is not a subclass of y.   QED

> *Division Thesis*: Reality divides exhaustively into individuals and classes; in other words, everything is an individual, a class, or a mixed fusion of individual and class.

Proof By our present definitions, that means that everything either has no singletons as parts, or is a fusion of singletons, or is a fusion of something with no singletons as parts and some singletons. We've already proved that everything either is a part of the null set, or is a fusion of singletons, or is a fusion of a part of the null set and some singletons. And by definitions and mereology, something is part of the null set iff it has no singletons as parts.   QED

> *Priority Thesis*: No class is part of any individual.

Proof That means that no fusion of singletons is part of anything that has no singletons as parts, which holds by mereology.   QED

> *Fusion Thesis*: Any fusion of individuals is an individual.

Proof That means that a fusion of things without singletons as parts itself has no singletons are parts, which holds by mereology plus the fact that singletons are atoms.   QED

These four theses, we recall, imply our

> *Main Thesis*: The parts of a class are all and only its subclasses.

4.4 Set Theory Regained

From the axioms and definitions of mereologized arithmetic, plus principles of the framework, we next derive versions of the standard axioms for iterative set theory. In some cases these versions will be stronger than usual, because schematic formulations will give way to plural quantification. In other cases they will be weaker than usual, because they will be conditional on our hypotheses about the size of Reality. Having affirmed the hypotheses, however, we are committed also to the unconditional versions.

> *Null Set*: The null set is a set with no members.

Proof It is a set by definition. It has no members because, by Distinctness, it has no singletons as parts.   QED

> *Extensionality*: No two classes have the same members; no class has the same members as the null set.

Proof If two classes had the same members, they would be two different fusions of the same singletons, contrary to Uniqueness of Composition. Every class has members, unlike the null set.   QED

> *Pair sets*: If each of x and y is an individual or a set, then there exists a set of x and y.

Set Theory Regained

Proof Each of x and y has a singleton, and the fusion of these singletons is the class of x and y. We have seen that singletons are finite and that they are small. Since there is something infinite that consists of atoms, the fusion of two singletons – something small and something finite – is small. The class of x and y is small, and hence a set. QED

> *Aussonderung*: Given a set x, and given some things, there is a set of all and only those of the given things that are members of x.

Proof If x is the null set, or if none of the given things are members of x, then the required set is the null set. Otherwise x is a small class; all members of x have singletons; so by Unrestricted Composition we have the fusion of all singletons of members of x that are among the given things. This is the class of all members of x that are among the given things. It is part of the small class x, hence small, hence a set. QED

Ordinarily, following Skolem, *Aussonderung* is stated as an axiom schema. This amounts to a special case of the plurally quantified version: given a set x, and given some things *that are all and only the satisfiers of such-and-such formula*, there is a set of all and only those of the given things that are members of x. (Or we may get the same effect in two stages: first, a version of *Aussonderung* that tells us that given a set x and a class y, there is a set of the common members of x and y; and second a class-comprehension schema to tell us that there is a class – perhaps proper – of all and only the sets and urelements that satisfy such-and-such formula.) There are normally restrictions, more or less stringent, on what the formula may be. Our plurally quantified *Aussonderung* is stronger. But beware: this is useless strength, when it comes to proving theorems in a formal system. For the formal system in question will have to include some axiom system, inevitably partial, for the framework. What we get out of our strong *Aussonderung* depends on what we can put into it. Recall that we can write principles of plural 'comprehension': if there is at least one thing that..., then there are some things that are all and only the things that.... We noted that such principles are ontologically innocent and altogether trivial. Nevertheless, an axiom system for the framework must supply them; and if, being incomplete, it fails to supply enough of them, that will cramp our style as surely as if *Aussonderung* itself had been given as a mere schema. So the point of using plural quantification here is not to increase our power to prove things, but rather to tell the whole truth.

The same comment will apply, *mutatis mutandis*, to plurally quantified versus schematic formulations of Replacement and Choice.

> *Replacement*: If there are some ordered pairs whereby each member of a class x is paired with exactly one member of a class y, and if for each member of y there is a member of x that is paired with it, and if x is a set, then y is a set.

We assume that pairing is defined here in some standard set-theoretical way, say that of Kuratowski: the (*ordered*) *pair* of u and v is the two-membered set whose members are the singleton of u and the set of u and v.

Proof Fill the blank in the Replacement Schema (singular version) with 'is such that its member is paired by one of the given pairs with the member of'. Recall that the atoms, i.e. the singleton subclasses, of x and y correspond one-one with their members, by Functionality and Distinctness. We conclude that if x is small then so is y. So y is a set if x is.   QED

Set Theory Regained

Fundierung: No class intersects each of its own members.

A counterexample to *Fundierung* might be an infinite class of $x_1, x_2, x_3, \ldots$, where x_1 contained x_2 as a member, and x_2 contained x_3, and so on; or a finite class of $x_1, x_2, \ldots, x_n$, where x_1 contained x_2, and ... and finally x_n contained x_1; or simply a single thing that was its own singleton. *Fundierung* says that none of these things can happen.

Proof Call something *grounded* iff it belongs to no class that is a counterexample to *Fundierung*. We show by Induction that everything is grounded, and conclude that *Fundierung* can have no counterexamples. First, an individual is grounded: it has no members, intersects nothing, and so intersects no class that contains it. Second, if y is grounded, so is its singleton; else the singleton belongs to, and hence intersects, a counterexample class C, in which case y must belong to C, *contra* the groundedness of y. Third, a fusion of grounded things is grounded. If the fusion is an individual, we already know that it is grounded. If it is a mixed fusion of individual and class, or a proper class, it is grounded because it isn't a member of anything. The remaining case is that the fusion is a small class, in which case the given grounded things are small classes too. Suppose for *reductio* that the fusion belongs to, and therefore intersects, a counterexample class C. Then one of the given grounded things, call it g, also intersects C. Let C^+ be the class that contains all the members of C, and g as well. Is C^+ another counterexample class? No, because it contains the grounded g. But yes, because each member of C^+ – the members of C, and g as well – intersects C^+ by intersecting C. This completes the *reductio*, and so completes the proof that everything is grounded.   QED

Choice: Suppose x is a class, and suppose there are some ordered pairs whereby each member of x is paired with at least one thing, and no two members of x are paired with the same thing. Then there is a class y such that each member of x is paired with exactly one member of y.

Proof Apply the First Choice Schema, filling in the blank with 'is paired by one of the given pairs with'. (Note that things which appear in pairs must have singletons, hence, whenever we have some of them, we have a class.)   QED

Power Sets: Given Hypothesis P about the size of Reality, then if x is a set, there is a set of all subsets of x.

Proof Since x is a set, x is small. Then the subclasses of x are also small; so they are sets, and they have singletons. The fusion y of these singletons is the class of all subclasses of x. The fusion z of y with the singleton of the null set – one further atom – is the class of all sub*sets* of z. By Hypothesis P, the parts of x – that is, its subclasses – are few. Then by the Replacement Schema (plural version) their singletons are few. These are the atoms of y, so y is small. We recall that since the fusion of all singletons is infinite and consists of atoms, the fusion of something small with one further atom, which is finite, is still small. So z is small, and therefore is a set.   QED

Unions: Given Hypothesis U about the size of Reality, then if x is a set, there is a set of all members of members of x.[3]

---

[3] It is not really necessary to take Unions in this conditional form. Azriel Lévy proved that unconditional Unions is redundant in von Neumann's set theory; see Lévy, 'On von Neumann's Axiom System for Set Theory', *American Mathematical Monthly*, 75 (1968), pp. 762–3. John P.

Set Theory Regained

Proof If all members of x are individuals (the null set or proper parts thereof) then the required set is the null set. Otherwise, consider those classes that are members of x. Let y be their fusion; then y is the class of members of members of x. Each of these classes, since it is a member of something,

---

Burgess has shown that an improved version of Lévy's proof, bypassing Lévy's use of unconditional Power Sets and therefore not relying on Hypothesis P, is available in the present system as well. We knew already that the axioms of mereologized arithmetic had implications concerning the size of Reality; Burgess's result indicates that they imply Hypothesis U.

Here is part, but only part, of a direct proof that mereologized arithmetic implies Hypothesis U. First, I conjecture that

> (★) If there are infinitely many atoms, and if some large thing is the fusion of a few small things, then Reality itself is the fusion of a few small things.

follows from evident principles of the framework. (I fear it is not a sufficiently evident principle in its own right. I believe it in virtue of a suspect analogy with what's true of things smaller than Reality.) Assume (★). Suppose for *reductio* that mereologized arithmetic holds but Hypothesis U is false. Then, since by mereologized arithmetic there are infinitely many atoms, we have by (★) that Reality is the fusion of a few small things. Then *anything* is the fusion of a few small things: namely, its intersections with some of the few small things whose fusion is Reality (whichever ones of them it overlaps). Given mereologized arithmetic, that enables us to code all things, and in particular all fusions of atoms, by single atoms; which is impossible, by the reasoning of Cantor's Theorem and Russell's Paradox.

The coding goes thus. First, we can code any x by a class y, such that if x is small, y is small too: if any singletons are parts of x, y contains them, and if any individuals are parts of x, y contains their fusion. If an arbitrary thing is the fusion of a few small things, we can code it by the few small classes that code those things, then by the few singletons of those classes, then by the fusion of those singletons, and finally by the singleton of that fusion, which is an atom. QED

must be small. Further, there must be few of them, since the class of them is part of the small class x. So y is the fusion of a few small things, hence small by Hypothesis U. Therefore y is a set.   QED

> *Infinity*: Given Hypothesis I about the size of Reality, then there is a nesting with no greatest member, the union of which is a set.

(A *nesting* is a class of sets, such that, whenever x and y are two members of it, either x is properly included in y or y is properly included in x.)

Proof By Hypothesis I, there are some things such that each of them is a proper part of another, and such that their fusion is a small fusion of stoms. Each of these things must be a small fusion of atoms. Each atom must be either an individual or a singleton, and in either case will have a singleton. Replace each atom by its singleton: that will not affect the smallness of the things and their fusion, nor will it affect the fact that each is a proper part of another.

Now we can say that there are some sets such that each of them is properly included in another, and such that their union is a set. Apply the Second Choice Schema, filling in the blank with 'is properly included in', and we may conclude that there are some of the given sets – the class of these will be the desired nesting – such that (1) each of them is properly included in another one, so that none is greatest; and (2) whenever x and y are two of them, either x is properly included in y or y is properly included in x. Further, since these sets are some of the previous sets whose union was a set, the union of these sets is part of a set, hence small, and hence a set.   QED

Ordinary Arithmetic and Mereologized Arithmetic

This is not the most familiar formulation of a set-theoretical axiom of infinity, to be sure; but with the set theory now at hand, we know that we can get to various more familiar versions.

What we have, given Hypothesis I, is a set that is infinite under one possible set-theoretical definition of infinity, and also under our original mereological definition. We already knew that we had an infinite *class* – namely, the fusion of all singletons, which is the large class of all sets and individuals – but not yet an infinite *set*. For all we knew without Hypothesis I, even given Hypotheses P and U, it might have been that all sets were finite, and only proper classes were infinite.

4.5 Ordinary Arithmetic and Mereologized Arithmetic

We could write the Peano axioms for the arithmetic of the natural numbers, with 'successor' as primitive, as follows.

Functionality: Nothing has two different successors.

Domain: Zero has a successor; any successor has a successor; and nothing else has a successor.

Distinctness: No two things have identical successors, nor is zero identical to any successor.

Induction: If there are some things, if zero is one of them, and if every successor of one of them is one of them, then everything is one of them.

In the axioms as stated, 'zero' looks like an unacknowledged second primitive. Let us set that right. Induction implies that everything is either zero or a successor; Distinctness says that zero is not a successor; hence it follows from the axioms that zero may be defined as the thing that is not a successor, and that there is such a thing. So the axioms could have been rewritten just in terms of 'successor'.

Four steps take us from ordinary to mereologized arithmetic. The first step is the crucial one: stipulate that besides getting new numbers from old ones by succession, we can also get new 'numbers' from old by fusion. Specifically, by taking small fusions of successors. The numbers so obtained, like other numbers, have successors of their own; these successors have their own successors, and they join in fusions; and so it goes. We have old-fashioned numbers 0, 1, 2, 3, . . .; and also the new 'number' that is the fusion of 1 and 3; and the successor of that fusion; amd the fusion of that successor with 2; and the successor of that second fusion. . . . Taking this step means adding clauses to two axioms. The extra clause for Domain is 'any small fusion of successors has a successor'. The extra clause for Induction is 'if every fusion of some of them is one of them'.

Why not: 'if every *small* fusion of some of them is one of them'? That would correspond directly to our new way of generating numbers. But when we say 'everything', do we mean *everything*, or do we just mean everything in some restricted domain? In the case of ordinary arithmetic, we had better just mean everything in a restricted domain – we're not Pythagoreans! But by the time we're done mereologizing arithmetic, we may hope to say 'everything' and mean it without restriction. Speaking unrestrictedly, there are large fusions – for instance, Reality – as well as small ones. So to be assured that we get everything, large or small, we must close under fusion generally.

The second step stipulates that the new numbers really are new. We wouldn't want it to turn out that 17 is really the fusion of 6 and 42; or that zero is really the fusion of 17 with the successor of the fusion of 9 and 6. So we amend Distinctness: in both its clauses, we replace distinctness in the sense of mere non-identity with distinctness in the strong sense of non-overlap.

The third step allows, in effect, that we may have not just one zero but many: many starting points that are neither successors nor fusions of successors. If we're quantifying unrestrictedly, now, we need to make a place for cats and quarks and points, and presumably these things are not successors or fusions thereof. We assume further that if there are many zeros, they are all and only the parts of one big zero. So we may reserve the proper name 'zero' henceforth for the one big zero, and speak of all the 'zeros' (including the big one) henceforth as 'parts of zero'. So we amend Domain, Distinctness, and Induction by writing 'any part of zero' instead of 'zero'. (Strictly speaking, we needn't amend Distinctness: not overlapping any part of zero is equivalent to not overlapping zero. But I give uniformity of the amendments precedence over simplicity of the final formulation.)

The final step, I claim, is not substantive but merely verbal: we write 'singleton' instead of 'successor', and 'null set' instead of 'zero', throughout.

4.6 What's in a Name?

Not substantive? – that will arouse some incredulity! Surely 'singleton' is one concept, 'successor' is another? But this is to overestimate how rich a conception we have of the member-singleton function and the number-successor function. All there is to our understanding in either case (apart from some *via negativa*, to the effect, for instance, that Possum is neither a singleton nor a successor) is just a formal *modus operandi*, given in one case by the axioms of mereologized arithmetic

Set Theory for Mereologists

and in the other case by the axioms of ordinary arithmetic. (Hence the appeal of structuralism.) And what's said about the *modus operandi* of 'successor' is that, within a certain restricted domain, it's just like that of 'singleton'. The axioms of ordinary arithmetic, which quantify restrictedly, are silent about what does or doesn't go on outside that restriction. They do *not* deny that things falling outside the restriction may have successors. For all that arithmetic tells us, maybe Possum has a successor, and Magpie has a successor, and maybe the fusion of Possum's successor and Magpie's has a successor, ... In short, there's nothing in our meager conceptions of the member-singleton function and the number-successor function to prove them different, and simplicity favours the hypothesis that they are the same. And if they are the same, then it matters not at all whether we say 'singleton' or whether we say 'successor'. A singleton by any other name would smell as sweet.

Suppose Gregson tells Holmes that a thief is plundering Piccadilly, and the police know nothing about him except for his distinctive *modus operandi*. In particular, it is not known whether he operates anywhere else. And Lestrade tells Holmes that a thief is looting all of London, and the police know nothing about him except for his distinctive *modus operandi*. And lo, the *modus operandi* of the London looter is just the same as that of the Piccadilly plunderer! Or rather, the *modus operandi* of the looter, when he's looting in Piccadilly-like places, is just like that of the plunderer. Now even Watson may arrive at the working hypothesis that these scoundrels are one and the same. And we would do well to follow his example.

I am not just saying, as Zermelo did and as structuralists do, that if we take the null set as zero and the singleton function as successor, then we get a set-theoretical model of arithmetic – one such model among many, the so-called 'Zermelo numbers'. And I am not just saying, as Quine might, that when we have many models we can choose one aribtrarily, and I happen to choose this one.

Rather, I am saying this. Assume, *pace* structuralism, that we should take 'singleton' as primitive. Then we must admit that somehow we have managed to give 'singleton' an unequivocal meaning. But if we did it for 'singleton', why not also for 'successor'? What's the good of being a part-time structuralist, for arithmetic but not for set theory? You get the worst of both worlds. You bear the burden of admitting that somehow, you know not how, such feats of meaning-giving are possible. Yet you still bear the burden of denying our naive conviction that 'successor' has in fact been given an unequivocal meaning. You might as well be hanged for two sheep as for one. And if we have somehow given meaning both to 'singleton' and 'successor', then the obvious hypothesis, what with the match of formal *modus operandi* and the lack of known differences, is that we gave them both the same meaning.

What sets Zermelo's modelling of arithmetic apart from von Neumann's and all the rest? It is Zermelo's that identifies the primitive of arithmetic with an appropriately *primitive* notion of set theory. Not, indeed, the notion usually chosen as primitive – but one that could and should have been. If there is something about certain meanings that especially suits them to become the meanings of primitive terms – and if there isn't, it's hard to understand how determinate meaning in thought and language is possible at all[4] – then it is to

[4] See David Lewis 'New Work for a Theory of Universals', *Australasian Journal of Philosophy*, 61 (1983), pp. 343–77; and 'Putnam's Paradox', *Australasian Journal of Philosophy*, 62 (1984), pp. 221–36.

Set Theory for Mereologists

be expected that, whenever possible given constraints on formal *modus operandi*, one primitive term should take on the meaning that already belongs to another primitive term.

4.7   Intermediate Systems

Between ordinary arithmetic and mereologized arithmetic there lie intermediate systems. One is *pure* mereologized arithmetic. Take only the first two of our four steps (plus the fourth if you like) to get

> *Functionality*: Nothing has two different successors.
>
> *Domain*: Zero has a successor; any successor has a successor; any small fusion of successors has a successor; and nothing else has a successor.
>
> *Distinctness*: No two things have overlapping successors, nor does zero overlap any successor.
>
> *Induction*: If there are some things, if zero is one of them, if every successor of one of them is one of them, and if every fusion of some of them is one of them, then everything is one of them.

This 'everything' is presumably said under a restriction; a restriction which omits most of the non-successors we think there are, but within which Unrestricted Composition holds. Applying our definitions of class, membership, and so on (written with 'successor' or 'singleton', it doesn't matter), we get pure set theory: set theory without urelements, in which the only memberless things are the null set, proper classes, and mixed fusions of the null set and classes.

The other intermediate system differs right at the beginning.

Completeness of Mereologized Arithmetic

We take fusions of numbers to get new numbers, but only when the numbers fused meet a special condition. Not only must it be a small fusion of successors; also, it must be an initial segment without a last term. I skip the definition, which exploits the resources of plural quantification to define precedence, and rest content with examples. One fusible segment consists of the numbers 1, 2, 3,...; its fusion is the new 'number' omega. Omega has a successor, omega + 1. The next fusible segment consists of the numbers 1, 2, 3,..., omega + 1, omega + 2, omega + 3,.... Its fusion is the number omega + omega. Then comes omega + omega + 1,.... And so on, as long as the fusions remain small. In short, we have the system of ordinal arithmetic; the ordinals are zero, successors, and limit ordinals, with each limit ordinal taken as the fusion of all preceding successors. We adjust the axioms of Domain and Induction to accommodate the limit ordinals and their successors, and strengthen Distinctness by putting non-overlap in place of non-identity.

If we embed ordinal arithmetic in full mereologized arithmetic, we find that although we are still following Zermelo's plan when it comes to successors – 'successor' means 'singleton' – we are following von Neumann's plan when it comes to limit ordinals. Omega, the fusion of 1, 2, 3,..., is the union of 0's singleton, 1's singleton, 2's singleton,...; which makes it the class of its predecessors 0, 1, 2,.... Likewise omega + omega is the fusion of 1, 2, 3,..., omega + 1, omega + 2, omega + 3, ...; which makes it the class of its predecessors 0, 1, 2, ..., omega, omega + 1, omega + 2,....

4.8 Completeness of Mereologized Arithmetic

Let's imagine a fictitious history of geometry: there never was a parallel *postulate*. Instead there was the parallel *conjecture*:

Set Theory for Mereologists

a perennial problem, never proved and never refuted. After millenia, someone proposed a diagnosis. 'The trouble may be,' he said, 'that the accepted axioms do not characterize the primitive geometrical notion of *congruence* well enough to settle the question. Perhaps there are two rival congruence relations, congruence$_1$ and congruence$_2$, both satisfying our accepted axioms of geometry' — not including any parallel postulate — 'and the reason we cannot settle the parallel conjecture is that it is true for congruence$_1$ but false for congruence$_2$.' This diagnosis would have been exactly right.

The case of arithmetic is different. What if someone proposed a parallel diagnosis of our failure to settle Goldbach's conjecture? 'The trouble may be that the accepted axioms do not characterize the primitive notion of *successor* well enough. Perhaps there are two rival successor relations, successor$_1$ and successor$_2$, both satisfying the Peano axioms; and the reason we cannot settle Goldbach's conjecture is that it is true for successor$_1$ and false for successor$_2$.' This time, the diagnosis is wrong. One reason why is that the Peano axioms characterize the successor relation up to isomorphism. I mean, of course, the Peano axioms in their full strength, not some elementary approximation thereof: Induction is to be stated by means of plural quantification. If we have two rival successor functions, both satisfying those axioms, they must be structurally alike. Each must be the image of the other under a one-one correspondence. That means that their difference cannot affect the truth of Goldbach's conjecture, or any other statement formulated in the language of arithmetic.[5]

Gödel has not been vanquished, of course. It is as true as

---

[5] See any discussion of the 'categoricity of second-order arithmetic', for instance in Robbin, *Mathematical Logic*, pp. 161–3.

Completeness of Mereologized Arithmetic

ever that any consistent formal system of arithmetic, even if it has all the Peano axioms, must omit some truths that ought to be theorems. But arithmetic is not to blame. The trouble, rather, is that the framework of plural quantification does not admit of any complete formalization. The distinctively arithmetical part of the system *is* complete.

Now, how about mereologized arithmetic? Is it like the case of geometry without a parallel postulate? Or is it like the case of arithmetic? Since mereologized arithmetic is set theory, we certainly have unsettled conjectures galore. And we found plenty of cause to complain that we lack any good understanding of the primitive member-singleton relation. We may well ask whether our incomplete understanding of the primitive notion of set theory bears some of the blame for our incomplete knowledge of what's true in set theory.

But in fact it doesn't. The case of mereologized arithmetic – that is, set theory; that is, mathematics – is like the case of ordinary arithmetic. Our philosophical understanding of the concept of a singleton is already good enough for purposes of mathematics. Better understanding would do nothing to remedy our mathematical ignorance. The trouble lies elsewhere.

Suppose we have two rival singleton functions, *singleton*$_1$ and *singleton*$_2$, both satisfying the stated axioms of mereologized arithmetic. We ask whether some unsettled conjecture in set theory might be true for singleton$_1$ and false for singleton$_2$.

No matter how mystified we may be about the nature of singletons, still we somehow seem to know that cats, for instance, or puddles or quarks or space-time points, are not singletons and have no singletons as parts. Even without any understanding of what singletons are, we seem to know somehow just how much of Reality is the singleton-free zone. That limits the difference between our two rival singleton functions: we may assume that they do not disagree about the demarcation between individuals and everything else. The individuals$_1$ are exactly the individuals$_2$. (As illustrated here, we mark defined terms with subscripts to indicate whether they are defined in terms of singleton$_1$ or singleton$_2$. To be an *individual*$_1$ is to have no singletons$_1$ as parts, to be an *individual*$_2$ is to have no singletons$_2$ as parts.) It follows, since singletons are exactly the atoms that are not individuals, that the singletons$_1$ are exactly the singletons$_2$. So we may suppress the subscripts when we classify things as 'individuals' and 'singletons'; but of course the subscripts must return when we speak of the singleton *of* something.

As well as the demarcation of the individuals, the framework also is settled. We are quantifying, once and for all, over everything; so the domain of quantification is fixed. The interpretation of the apparatus of the framework, plural quantification and mereology as well as elementary logic, is fixed. That means that singleton$_1$ and singleton$_2$ cannot differ in what they make true concerning the size of Reality. That is a framework matter, to which the interpretation of 'singleton' is irrelevant.[6]

Then, I say, singleton$_1$ and singleton$_2$ differ only by a permutation of singletons. They are structurally alike. So they cannot differ in any way that matters to the truth of any statement in the language of mereologized arithmetic. Given the fixed demarcation of the individuals, given the fixed framework and consequently the fixed size of Reality, the

---

[6] Here it is important that we reject the singularist dogma that plural quantification must really be singular quantification over classes, or at any rate over some sort of class-like entities. A singularist has no business holding the interpretation of plural quantification fixed while regarding the interpretation of the set-theoretical primitive as unsettled. See Thomas Weston, 'Kreisel, the Continuum Hypothesis, and Second Order Set Theory', *Journal of Philosophical Logic*, 5 (1976), pp. 281–98.

Completeness of Mereologized Arithmetic

axioms of mereologized arithmetic suffice to characterize the primitive notion of singleton up to isomorphism – indeed, up to *auto*morphism.[7]

It remains to state this properly, and to prove it. A *relation of singletons* is a class, perhaps proper, of ordered pairs of singletons.[8] It is a *permutation of singletons* iff each singleton is the first term of exactly one pair and the second term of exactly one pair. If P is a permutation of singletons, singleton$_1$ and singleton$_2$ *differ by* the permutation P iff

(1) whenever x is an individual, P pairs the singleton$_1$ of x with the singleton$_2$ of x; and (2) whenever x and y are fusions of singletons, and P pairs each singleton in x

---

[7] There are known results concerning the *almost*-categoricity of second-order set theory: one model is not necessarily isomorphic to another, but at least one is isomorphic to an initial segment of the other. These are discussed, for instance, in Weston, 'Kreisel, the Continuum Hypothesis, and Second Order Set Theory'. It is our holding fixed of the framework that takes us from almost-categoricity to categoricity *simpliciter*.

If we had declined to hold fixed the demarcation of individuals, then the rival singleton functions could have differed structurally. Something that is a singleton$_1$ could also have been an individual$_2$ atom, or *vice versa*. So the two rival singleton functions might have disagreed about how many individual atoms there were, and in this way fallen short of isomorphism. The results concerning almost-categoricity avoid such problems by confining themselves to pure set theory; but we cannot take this course if we want to hold the domain of quantification fixed by quantifying unrestrictedly over everything.

[8] In section 2.6, I asked how we were entitled to speak of ordered pairs before we were given, unequivocally, some singleton function. So far (but see the appendix) we can introduce pairing only by a set-theoretic definition. Then how can we speak of ordered pairs now? – Because now I have supposed that we *are* given, unequivocally, each of the two rival singleton functions. Our problem is not privation but overabundance. We have only to make an arbitrary choice: as it might be, that our ordered pairs shall be Kuratowski pairs$_1$.

Set Theory for Mereologists

with a singleton in y, and for each singleton in y, P pairs some singleton in x with it, and x has a singleton$_1$, and y has a singleton$_2$, then P pairs the singleton$_1$ of x with the singleton$_2$ of y.

We define P to be the least relation that satisfies the closure conditions (1) and (2). So, whenever P pairs one thing with another, it must do so either in virtue of clause (1) or in virtue of clause (2). It remains to prove that P is a permutation of singletons.

Proof Call a singleton *good* iff P pairs it with exactly one singleton; and, in general, call something *good* iff every singleton that is part of it is good. We use Induction for singleton$_1$ to show that everything is good. Individuals are good, vacuously, because they have no singletons as parts. Fusions of good things are good. It remains to show that if x is good and y is x's singleton$_1$, then y is also good. Given that x has a singleton$_1$, x cannot be either a large fusion of singletons or a mixed fusion. We have two remaining cases.

Case 1: If x is an individual, P pairs y with x's singleton$_2$ by clause (1); and with nothing else, since an individual cannot be a fusion of singletons as well, and hence clause (2) cannot apply to x.

Case 2: If x is a small fusion of singletons, then let z be the fusion of all singletons which P pairs with singletons that are parts of x. Since x is good, the singletons that are part of it are good; by that and Replacement, we have that z also is small. So z has a singleton$_2$, call it w, and P pairs y with w by clause (2). Now suppose for *reductio* that P also pairs y with v, $v \neq w$. Since v is a singleton, let it be the singleton$_2$ of u, and $u \neq z$. How did y and v get paired? Not by clause (1): x is a fusion of singletons, hence not an individual. By clause (2),

Completeness of Mereologized Arithmetic

then. So u is a fusion of singletons, and P pairs each singleton in x with a singleton in u, and for each singleton in u, P pairs some singleton in x with it. It follows, by definition of z and goodness of the singletons in x, that $u = z$. Contradiction.

This completes the inductive proof that everything is good, and in particular that every singleton is good. That is: for every singleton, P pairs it with exactly one singleton. In just the same way, except that we use Induction now for singleton$_2$, we show that, for every singleton, P pairs exactly one singleton with it. Therefore P is a permutation of singletons. QED

To see better what it means that singleton$_1$ and singleton$_2$ differ by a permutation of singletons in the sense given by clauses (1) and (2), it is helpful to note that, in a derivative sense, P permutes not only the singletons but everything else as well. Say that *P maps x to y* iff y is the fusion of all individuals that are part of x, together with all singletons that P pairs with singletons that are part of x. It is easy to see, once we know that P is a permutation of the singletons, that P maps each thing to one and only one thing, and that, for each thing, P maps something to it.

Mereological operations are preserved: if P maps x to y, and z to w, then x is part of z iff y is part of w. There are two cases in which x has a singleton$_1$: when it is an individual and when it is a small class. In the second case, if P maps x to y, y also must be a small class and therefore must have a singleton$_2$. Clause (1) says that in the first case, when P maps x to itself, then P maps x's singleton$_1$ to x's singleton$_2$. Clause (2) says that in the second case, if P maps x to y, then P maps x's singleton$_1$ to y's singleton$_2$. So, in general, if P maps x to y, then P maps the singleton$_1$ of x (if such there be) to the singleton$_2$ of y. The sense in which singleton$_1$ and singleton$_2$ are structurally alike is, loosely speaking, that singleton$_2$ is

Set Theory for Mereologists

the image of singleton$_1$ under a mereology-preserving permutation of everything.[9]

That is why no statement in the language of mereologized arithmetic can be true for singleton$_1$ and false for singleton$_2$. No choice between the two rivals could tell us more than we know already about what's true in set theory. The sources of mathematical ignorance lie elsewhere: in our ignorance of the size of Reality, and in our irremediable lack of a complete axiom system for plural quantification.

---

[9] This is loose speaking because there can be no such thing as a permutation of everything. Not even a proper class can have as many members as there are pairs of things. That is why, hitherto, I shunned the noun and preferred the verb: *P* permutes everything.

There are ways, however, to *simulate* quantification over such relations as a permutation of everything (see the appendix).

Appendix on Pairing

by John P. Burgess, A. P. Hazen, and David Lewis

Introduction

Suppose we have the resources of plural quantification and mereology, as in chapter 3; but no primitive singleton function. Can we somehow get the effect of quantifying over relations? Equivalently, can we get the effect of plural quantification over ordered 'tuples? We can, as it turns out, if we have infinitely many atoms, and not too much atomless gunk. It can be done in two different ways, or three if we count one that is just a hybrid of the other-two.

Until further notice, we shall assume that everything consists entirely of atoms. Later we shall extend both methods to cover the case where there is a certain amount of atomless gunk as well.

The Method of Double Images (Burgess, 1989)

The foremost 'paradox of infinity' is that the part may equal the whole. Infinite Reality contains infinite proper parts that are microcosms, each one containing within itself images of

Appendix on Pairing

all the things there are. Suppose we have one such microcosm; and also a second, wholly distinct from the first.[1] All things have images within the first microcosm, call them *first images*; and all things have images within the second, call them *second images*. Any first image is distinct from any second image. Then if p_1 is Possum's first image and m_2 is Magpie's second image, the fusion $p_1 + m_2$ unequivocally codes the pair of Possum and Magpie, taken in that order. Such fusions of first and second images may serve in general as ordered pairs.

To make good sense of the plan just sketched, we must first explain what we mean by our talk of 'images'. We cannot yet say directly that there exists a one-one mapping from all atoms into the atoms of the first microcosm, and there exists another one-one mapping from all the atoms into the atoms of the second microcosm. Discovering how to say such things is exactly our present task.

But we *can* speak, in effect, of one-one mappings between the atoms of *distinct* things. In this special case, we can specify a relation without recourse to ordered pairs. We can use *unordered* pairs of atoms: two-atom fusions, or *diatoms*. We quantify plurally over them.[2]

> *Definition*: Diatoms Y map x one-one into y iff x and y are distinct; any two of Y are distinct; each one of Y consists of an atom of x and an atom of y; and x is part of the fusion of Y.

---

[1] We follow Lewis's usage: 'distinct' means 'non-overlapping' or 'disjoint', rather than 'non-identical'. (See section 3.4.)

[2] Soon we shall meet strings of several plural quantifiers, so to avoid confusion we supplement the plural pronouns of English with capital letters used as plural variables. Sometimes we also use capital letters as singular variables over relations; for all such variables will turn plural upon translation.

The Method of Double Images

We cannot tell the direction of mapping just by looking at a diatom, but since x and y are given along with the diatoms Y, we know that we are to take the diatom as mapping the atom of x to the atom of y and not *vice versa*. The stipulation that x and y are distinct makes sure we will never be confused by atoms that are common to both.

So far, we are limited to the special case that the domain and range are distinct. But we can get around this limitation by taking mappings in tandem, as follows. Suppose that x_1 and x_2 are a *partition* of x: that is, they are distinct, and their fusion is x. Suppose likewise that y_1 and y_2 are a partition of y. Suppose that the diatoms Y_1 map x_1 one-one into y_1, and the diatoms Y_2 map x_2 one-one into y_2. Then Y_1 from x_1 to y_1, in tandem with Y_2 from x_2 to y_2, together map x one-one into y. Note that we have written our tandem mapping as an eight-place relation (with two places plural). We have not just merged the diatoms Y_1 and the diatoms Y_2. We have retained the information that Y_1 are to be taken as mapping from x_1 to y_1 and Y_2 are to be taken as mapping from x_2 to y_2. The limitation to distinct domains still applies to x_1 and y_1; it still applies to x_2 and y_2. But it does not apply to x and y. They may overlap. They may even be identical, as shown in figure 1: $x_1 = y_2$, $x_2 = y_1$, so $x = y$.

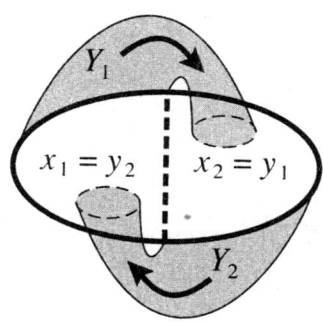

Figure 1

Appendix on Pairing

By taking mappings in tandem, twice over, we can map the whole of Reality one-one into each of two distinct microcosms. We first affirm the following as a framework principle. (It might be derived from other, more evident principles; but since we can have no complete axiom system for the framework, and we have not chosen any particular partial axiom system, we refrain from doing so.)

> *Trisection*: If Reality is infinite, and consists entirely of atoms, then there exist some x, y, z, and some X, Y, Z, such that x, y, and z partition Reality; X maps $y+z$ one-one into x; Y maps $x+z$ one-one into y; and Z maps $x+y$ one-one into z.

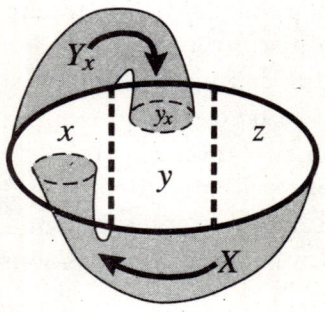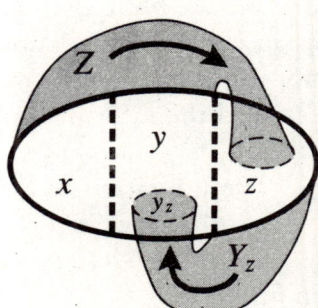

Figure 2

Given any such x, y, z, X, Y, Z, we proceed as shown in figure 2. First, we divide Y and y. Let Y_x be those diatoms of Y that have an atom of x as part and let y_x be the fusion of those atoms of y that are part of diatoms of Y_x; then the diatoms Y_x map x one-one into y_x. Define Y_z and y_z in the same way; the diatoms Y_z map z one-one into y_z. Now X from $z+y$ to x, in tandem with Y_x from x to y_x, together

The Method of Double Images

map the whole of Reality one-one into $x+y_x$. So $x+y_x$ contains our first microcosm. Likewise Z from $x+y$ to z, in tandem with Y_z from z to y_z, together map the whole of Reality one-one into $z+y_z$. So $z+y_z$ contains our second microcosm. The two microcosms are distinct.

Now we can make good our talk of images. We start with images of atoms, and go on to images of fusions of atoms. All imaging is relative to a given x, y, z, X, Y, Z, assumed to satisfy the conditions stated in the principle of Trisection.

> *Definition*: Atom v is the *first image* (with respect to x, y, z, X, Y, Z) of atom u iff either the diatom $u+v$ is one of X and u is part of y or of z, or else the diatom $u+v$ is one of Y and u is part of x. The *first image* (with respect to...) of any fusion of atoms is the fusion of the first images of its atoms.

Likewise, *mutatis mutandis*, for the definition of second images. Any fusion of atoms has a unique first image and a unique second image. All first images are parts of $x+y_x$ and all second images are parts of $z+y_z$. Hence any first image is distinct from any second image. Finally, no two different things ever have the same first or second image. Now we define our ordered pairs.

> *Definition*: The b-pair (with respect to...) of f and s is the fusion of the first image of f and the second image of s.

For any f and s, there is a b-pair of f and s. A b-pair is unambiguous: if p is the b-pair of f and s, and if p' is the b-pair of f' and s', then $p=p'$ only if $f=f'$ and $s=s'$.

By repeated pairing, we have ordered 'tuples of other

Appendix on Pairing

degrees. Let the *b-triple* of f, s, and t be the b-pair of f and p, where p in turn is the b-pair of s and t. (We must be told, of course, whether such a thing is meant to be taken as a pair or as a triple.) Similarly we have ordered quadruples, quintuples, ...

Roughly speaking, a singular quantifier over relations may be replaced by a plural quantifier over ordered 'tuples of the appropriate degree. Instead of saying 'there is a (dyadic) relation', say 'there are some b-pairs'; and similarly for triadic, tetradic, . . . relations.

But only roughly. The complication is that our b-pairs do not have the wherewithal for decoding built into them. In this respect they are inferior to the set-theoretical ordered pairs of Wiener or Kuratowski. A quantifier over b-pairs must therefore be accompanied by quantifiers over the wherewithal for decoding. If we have a singular existential quantifier over dyadic relations, and within its scope one or more atomic formulas,

> For some relation R: ——— jRk ———,

then our translation must take the cumbersome form

> For some x, y, z, X, Y, Z such that [here follow the conditions stated in Trisection], for some R such that each one of R is a b-pair (with respect to x, y, z, X, Y, Z): ——— the b-pair of j and k (with respect to x, y, z, X, Y, Z) is one of R ———

with six extra initial quantifiers besides the plural quantifier over pairs.[3] Similarly, if we translate a singular universal

---

[3] The order of initial existential quantifiers does not matter; if we swap the x and z quantifiers, and swap the X and Z quantifiers, we get a

The Method of Extraneous Ordering

quantifier over dyadic relations, we get a string of seven universal quantifiers.

The same complication will arise, *mutatis mutandis*, for all other methods of pairing considered in this appendix, though the wherewithal for decoding the pairs will look different in different cases. It is to be understood, therefore, that any plural quantifier over pairs will be accompanied by a string of other quantifiers, mostly plural. For the most part we shall leave these accompaniments tacit.

The Method of Extraneous Ordering (Hazen, 1989)

There is a second special case in which we can specify a relation without recourse to ordered pairs. Take the parent-child relation among people. If we know which people are older than which, we can specify the parent-child relation by means of the *unordered* parent-child pairs. The parent is always the older one. We lack the built-in ordering whereby the first term of an ordered pair precedes the second, but the

---

sentence equivalent to the original translation. But the b-pair of s and f with respect to x, y, z, X, Y, Z is the same thing as the b-pair of f and s with respect to z, y, x, Z, Y, X. Does this mean that, after all, we have lost track of the direction of the pairs and the relation? – Not in any sense that matters. Consider sentences

(1) For some R: ——jRk ——mRn——
(2) For some R: ——kRj ——nRm——
(3) For some R: ——jRk ——nRm——

with no occurrences of 'R' except those in the displayed atomic formulas. To be sure, our translations of (1) and (2) are equivalent. So they should be. The original (1) and (2) also are equivalent: whenever there is a relation, there is also its converse. What matters is that the translations of (1) and (3) are not equivalent. Within the scope of the quantifier, we still keep track of sameness and difference in the direction of the relation.

Appendix on Pairing

extraneous ordering by age serves as a substitute.

The relation of respect among people is more of a problem. Given an unordered pair of two people such that one respects the other, we cannot tell whether it is the older who respects the younger, the younger who respects the older, or whether the respect is mutual. Also, some people respect themselves and some do not.[4] However, we can specify the relation of respect by dividing it into three subrelations. First, there is the relation of respect from older to younger; it can be given by its unordered pairs, just as the parent-child relation can. Second, there is the relation of respect from younger to older, and it too can be given by its unordered pairs. Third, there is the relation of self-respect; it can be given by its unordered identity-pairs. More simply, it can be given just by giving its 1-tuples, the self-respecting people themselves.[5]

The case of a triadic relation R among people is cumbersome, but no different in principle. We divide it into thirteen subrelations: six triadic, six dyadic, and one monadic. They are given as follows. (We write 'O' for 'is older than'.)

---

[4] What of the case that someone respects someone else of precisely the same age? We may assume, at safe odds of infinity to one, that this never happens – not when the ordering by age has been specified in a sufficiently precise (and artificial) fashion.

[5] We could instead divide respect into just two subrelations: the relation of respect for someone not older than oneself, and the relation of respect for someone not younger than oneself. Each subrelation is given by its unordered pairs; however, some of these 'pairs' will be identity pairs of self-respecting persons, or 1-tuples, or simply the self-respecting persons themselves. This variation would be easier if we had to do only with dyadic relations, but when we go on to triadic relations it is much more trouble than it is worth.

The Method of Extraneous Ordering

$R_1 xyz$ iff $Rxyz$ & $xOyOz$ $R_7 xy$ iff $Rxxy$ & xOy
$R_2 xyz$ iff $Rxyz$ & $xOzOy$ $R_8 xy$ iff $Rxxy$ & yOx
$R_3 xyz$ iff $Rxyz$ & $yOxOz$ $R_9 xy$ iff $Rxyy$ & xOy
$R_4 xyz$ iff $Rxyz$ & $yOzOx$ $R_{10} xy$ iff $Rxyy$ & yOx
$R_5 xyz$ iff $Rxyz$ & $zOxOy$ $R_{11} xy$ iff $Rxyx$ & xOy
$R_6 xyz$ iff $Rxyz$ & $zOyOx$ $R_{12} xy$ iff $Rxyx$ & yOx
$R_{13} x$ iff $Rxxx$

Each of $R_1 \ldots R_6$ can be given by its unordered triples; each of $R_7 \ldots R_{12}$ by its unordered pairs; and R_{13} by its 1-tuples.

To what extent can we use this strategy within the framework of plural quantification and mereology, without benefit of set theory? So far, without benefit of set theory, we do not even have unordered pairs or triples of arbitrary things. People are one special case, because a fusion of people divides into people in only one way. Atoms are another special case, for the same reason. (A more important case, since more of Reality consists of atoms than consists of people.) Diatoms serve as unordered pairs. Likewise three-atom fusions, *triatoms*, serve as unordered triples; and atoms themselves serve as 1-tuples.

In the case of the atoms, unlike the case of people, we don't know how to specify an extraneous ordering. But we don't have to. It is enough to say that one exists. Then we can proceed relative to an unspecified value of the quantified variable. Or rather, values; for the quantified variable is plural. In the case of atoms (or in the case of people, but not in general) we can give an ordering by giving all fusions of its initial segments.

> *Definition*: O are *nested* iff, for any two of O, one of them is part of the other.

Appendix on Pairing

Definition: Atom *x precedes* atom *y* with respect to O iff there is some one of O that contains *x* but not *y*.

Definition: O *order* the atoms iff O are nested; the fusion of O contains all atoms; and for any two atoms, one precedes the others with respect to O.

We might affirm, as a principle of the framework, the existence of some O which order the atoms. Instead we affirm a stronger principle, a version of the Axiom of Choice (different from those stated in section 3.3), from which that follows.

Definition: O *well-order* the atoms iff O order the atoms and, whenever there are some atoms X, some one of X precedes all the others with respect to O.

Well-ordering Principle: Some O well-order the atoms.

Now we can translate a quantifier over relations of atoms. For the dyadic case, we translate

For some relation R: ——jRk——,

as

For some O that order the atoms, for some diatoms R_1, for some diatoms R_2, for some atoms R_3: —— either $j+k$ is one of R_1 and j precedes k with respect to O, or $j+k$ is one of R_2 and k precedes j with respect to O, or $j=k$ and j is one of R_3——;

The Method of Extraneous Ordering

and similarly when the quantifier is universal.[6] For the triadic case we have a string of fourteen plural quantifiers: first comes the O quantifier, then the $R_1 \ldots R_6$ quantifiers over triatoms, then the $R_7 \ldots R_{12}$ quantifiers over diatoms, and finally the R_{13} quantifier over atoms. An atomic formula 'Rhjk' within the scope of the quantifier becomes a disjunction with thirteen disjuncts.

We spare you the tetradic case; because once we can quantify over triadic relations, we can introduce ordered pairs of atoms. Then we can proceed to ordered triples, quadruples, ... of atoms by repeated pairing; and we can translate all further quantifiers over relations among atoms by plural quantifiers over 'tuples.

Our ordered pairs of atoms will themselves be atoms. But we need not specify our pairing relation; it is enough to say that one exists.[7] But we do not say exactly that; rather, we say its translation. The requisite definition and principle are given below in untranslated form, with a quantifier over triadic relations of atoms. What we really affirm as a principle of the framework is the translation of the principle as written.

> *Definitions*: P is a *pairing relation* (on atoms) iff P is a triadic relation of atoms; for any atoms f and s, there is a

---

[6] One might worry that we lose track of the direction of the relation, since it would be reversed if we reversed the ordering O or if we swapped the R_1 and R_2 quantifiers. But the reply given in note 3 applies, *mutatis mutandis*.

[7] Instead of our one triadic pairing relation, we could take two dyadic *un*pairing relations: the one that yields the first term, and the one that yields the second term. Together, these give back the pairing relation. Had we chosen this alternative, it would have been worthwhile to adopt the variation considered in note 5.

Appendix on Pairing

unique p such that $Pfsp$; if $Pfsp$ and $Pf's'p$, then $f=f'$ and $s=s'$. The *h-pair* of f and s (with respect to P) is the p such that $Rfsp$.

Pairing: If Reality is infinite, and consists entirely of atoms, there exists a pairing relation.

Now our new translation of

For some relation R: ——jRk——,

is our old translation of

For some h-pairs R, with respect to some pairing relation P: ——the h-pair of j and k (with respect to P) is one of R——;

and similarly when the quantifier is universal; and similarly in the triadic, tetradic,... cases. We need only use an extraneous ordering once, to legitimize the introduction of h-pairs by quantifying over pairing relations. Thereafter we use the h-pairs.

So far, we have a pairing relation that only yields h-pairs of single atoms. But we may extend the definition of h-pairing from atoms to fusions. Given two fusions f and s of atoms, we may pair them atom by atom, as follows.[8]

Definition: The *h-pair* of f and s (with respect to P) is the fusion of all h-pairs (with respect to P) of an atom of f and an atom of s.

---

[8] Here we borrow a device that is known in other uses, for instance in higher-order arithmetic. We thank W. V. Quine for pointing out to us that it would serve our present purpose.

Adding Gunk

Since h-pairs of atoms are atoms, we can recover them from their fusion. Then we can recover f as the fusion of the atoms which are the first terms of these h-pairs, and s as the fusion of the atoms which are the second terms. So h-pairing of fusions is unambiguous.

We obtain ordered 'tuples of other degrees by repeated h-pairing. We can translate a singular quantifier over relations between fusions of atoms by a plural quantifier (suitably accompanied) over such 'tuples.

A Hybrid Method

We could begin as the Method of Extraneous Ordering does, and finish as the Method of Double Images does. Once we can quantify over dyadic relations of atoms, however we do it, we can state a framework principle as follows: If Reality is infinite, and consists entirely of atoms, then there exist two relations R_1 and R_2 mapping Reality one-one into two distinct proper parts of Reality. With respect to any such R_1 and R_2, we can define the *first* and *second images* of a given thing as its images under R_1 and R_2 respectively. We can define the *hb-pair* of f and s as the fusion of the first image of f and the second image of s. We go on as usual to 'tuples of higher degree and to a translation of singular quantifiers over relations.

Adding Gunk

Now we withdraw the assumption that Reality consists entirely of atoms. Maybe the fusion of all atoms is only part of Reality; and the rest is atomless gunk, so that any part of it has proper parts in turn. If so, all our methods so far

Appendix on Pairing

apply only to the atomic part of Reality. They must all be extended.

To that end, we note still another special case in which a relation may be given by unordered pairs: the case of a relation that holds between atomless things and atoms. The relation is given by the gunk-and-one-atom fusions such that the gunk bears the relation to the atom. In general, any fusion decomposes uniquely into its gunk and its atoms.

> *Definition*: The *maximal atomless part* and the *maximal atomic part* of any given thing are, respectively, the fusion of all its atomless parts, if any, and the fusion of all its atoms, if any.

A gunk-and-one-atom fusion has maximal parts of both kinds, and the maximal atomic part is a single atom. Now we want to say, by plural quantification over gunk-and-one-atom fusions, that there is a one-one correspondence whereby every atomless thing has a representative atom. That will be so, provided there are enough atoms to go around; in other words, provided there is not too much atomless gunk.

> *Not-too-much-gunk Hypothesis*: There are some things G such that each one of G is the fusion of an atomless thing and exactly one atom; every atomless thing is the maximal atomless part of exactly one of G; and no two of G have an atom as a common part.

(The hypothesis follows from Hypothesis P of section 4.1. The fusion of all atomless things is small; so by Hypothesis P its parts are few; these parts are all and only the atomless things; and by the definition of 'few', we have the desired G.)

Let $A(x)$ be the maximal atomic part of x, if x has any atoms. And, given G satisfying the conditions of the

Adding Gunk

Not-too-much-gunk Hypothesis, let $G(x)$ be the atom such that its fusion with the maximal atomless part of x is one of G, if x has any atomless parts.

We have our methods of introducing ordered pairs, triples, etc, within the atomic part of Reality. Whichever method we choose, our pairs, triples, etc. are fusions of atoms. Choose some one of our methods of pairing, and hold it fixed in what follows.

We can encode arbitary things by ordered pairs of fusions of atoms. Something x may consist entirely of atoms, so that $G(x)$ is undefined; or it may be atomless gunk, so that $A(x)$ is undefined; or it may be mixed, part atoms and part atomless. Let a, g, m be three arbitrarily chosen atoms, used as markers to distinguish the three cases.

> *Definition*: The *code* of x (with respect to G, a, g, m) is the pair of a and $A(x)$, if $G(x)$ is undefined; or the pair of g and $G(x)$, if $A(x)$ is undefined; or the pair of m and the pair of $A(x)$ and $G(x)$, otherwise.

Now we can introduce new ordered pairs, available not only for fusions of atoms but for things of all kinds, as ordered pairs of these codes.

> *Definition*: The *new pair* of f and s is the pair of the code of f and the code of s.

We obtain new triples, quadruples,...by repeated pairing. We translate a singular quantifier over relations between arbitrary things into a plural quantifier over the new 'tuples. (The accompanying quantifiers over the wherewithal for decoding the new 'tuples now include quantifiers over G, a, g, m, as above, as well as the wherewithal for decoding the old pairs.) Henceforth we may quantify over relations, freely

Appendix on Pairing

and without comment. Knowing, as we do, how such quantifiers may be translated, we need not bother actually to translate them.

Megethology[9]

Sections 3.7 and 4.1 used the resources of the framework to express distinctions and hypotheses concerning the size of Reality. Now that the resources of the framework turn out to include quantification over relations, the job can be done over in an easier and more familiar way.

> *Definition*: X and Y are *equinumerous* iff there exists a relation R such that each one of X bears R to exactly one of Y, and for each one of Y, exactly one of X bears R to it.
>
> *Definitions*: Y are *many parts of* x iff Y are parts of x, and the atoms of x and some of Y are equinumerous. Y are *few parts of* x iff Y are parts of x but not many parts of x.
>
> *Definitions*: Something is a *large part of* x iff its atoms are many parts of x; a *small part of* x iff its atoms are few parts of x.

Corresponding to Hypotheses P, U, and I of section 4.1, we can write down three conditions on an arbitrary thing x. Taken together, they comprise a definition of inaccessible size.

> *Condition P*: Whenever something is a small part of x, its parts are few parts of x.

---

[9] From *megethos* 'size' + *logos* 'doctrine'.

Condition U: Whenever some things are few parts of x and small parts of x, their fusion is a small part of x.

Condition I: Some fusion of atoms of x is infinite and yet is a small part of x. (Something infinite is called *uncountable* iff it satisfies Condition I, otherwise *countable*.)

Definition: x is *of inaccessible size* iff x satisfies Conditions P, U, and I.

Standard set theory requires that Reality be of inaccessible size. But we may speculate, if we wish, that it is larger still.

Hypothesis IC: Some small part of Reality is of inaccessible size.

This is the framework version of a modest 'large cardinal' axiom. Given mereologized arithmetic, it implies that there is a set of inaccessible size, hence that there exists an inaccessible cardinal.

We can also state a less modest hypothesis, as follows. Suppose we have a measure, for instance the familiar Lebesgue measure of area. We can then distinguish regions of measure zero, 'negligible' regions, from others. Sometimes this qualitative distinction is all we want from a measure, and when it is, we can conflate the measure itself with the distinction between negligible regions and others. Measures in this purely qualitative sense can be characterized in the framework.

Definition: N measure x iff the fusion of N is x; x is not one of N; every part of one of N is itself one of N; and whenever y and z are among N, then the fusion of y and z is one of N.

Appendix on Pairing

When N measure x, N may also meet further conditions.

Defintion: N are *fully additive* on x iff, whenever Y are few parts of x and Y are among N, then the fusion of Y is one of N.

Definition: N are *maximal* on x iff, whenever y and z partition x, either y or z is one of N.

Definition: x is *measurable* iff, for some measure N on x, N are both fully additive and maximal on x.

A countable fusion of atoms is measurable.[10] Are larger things ever measurable? Not unless they are very much larger, as it turns out. An uncountable measurable thing would have to be of more than merely inaccessible size. But we may speculate, if we wish, that Reality is measurable. Or, stronger still,

Hypothesis MC: Some small part of Reality is measurable but uncountable.

Given mereologized arithmetic, this hypothesis implies that there exists a measurable cardinal.[11]

---

[10] If x is an infinite fusion of atoms, countable or not, the finite parts of x measure x. Whenever N measure x but fail to satisfy maximality, then N may be extended to N^+ which also measure x and do satisfy maximality; this is shown by an argument requiring a well-ordering of the parts of x. (It can be carried out when we have set theory, or else it requires additional principles of the framework.) Finally, we note that when x is countable, any measure on x must be fully additive; for in that case 'few' means 'finite', so full additivity adds nothing to the final clause in the original definition of measure.

[11] For a set-theoretical discussion of measurable cardinals, see Shoenfield, *Mathematical Logic*, section 9.10.

Ramsifying out the Singleton Function

As well as asking what is the largest size Reality has to offer, we can ask also about intermediate sizes. The Generalized Continuum Hypothesis, which addresses this question, is usually taken as a hypothesis about the sets. But we can state it in the framework.

> *Generalized Continuum Hypothesis*: If x is an infinite fusion of atoms, and Y are some of the parts of x, then either Y and all the parts of x are equinumerous or else Y and some of the atoms of x are equinumerous.

The Continuum Hypothesis, ungeneralized, is the special case in which we stipulate that x is countable.

Ramsifying out the Singleton Function, Continued

Now that we may quantify over relations, secure in our knowledge of how to translate these quantifications using only the resources of the framework, we may return to the unfinished business of section 2.6. Adapting a suggestion of Paul Fitzgerald, we considered a way to get rid of the primitive member-singleton relation. We could Ramsify it out. Take some axioms, perhaps the axioms of mereologized arithmetic in section 4.2, with 'singleton' as their only set-theoretical primitive.[12] Conjoin them into a sentence saying that the member-singleton relation satisfies certain structural (and perhaps other) conditions.

...singleton...singleton...singleton...

---

[12] To put them in primitive notation, uniformly replace 'null set' by 'fusion of all things that have no singletons as parts'; then uniformly replace 'singleton', when used non-relationally, by 'thing that is a singleton of something'; and add an extra axiom saying that something has no singletons as parts.

Appendix on Pairing

We may replace it by (a translation of) its Ramsey sentence

 For some S: ...S...S...S...

which says just that some relation satisfies those conditions.[13]

 Where there is one suitable relation, there will be many. (As many as there are permutations of the singletons.) To claim a primitive understanding of the member-singleton relation is to think that one of all these relations, and only one, is the one we had in mind all along. To retreat to set-theoretical structuralism[14] is to think that the suitable relations are all on a par, with nothing to distinguish the one real member-singleton relation from the horde of pretenders. In that case, any further sentence about 'the' member-singleton relation

 ——singleton——singleton——singleton——

should be understood as tacitly general: it says that something holds for all relations that satisfy the given conditions.

---

[13] Had we had been content to take membership generally as a relation between atoms, we could have Ramsified membership as soon as we were in a position to quantify over dyadic relations of atoms; we would have needed only the first part of the Method of Extraneous Ordering. The axioms to be Ramsified would not have been those of mereologized arithmetic, but would have taken membership as primitive in the standard way. This would have come closer to Fitzgerald's original suggestion. But it would have meant abandoning the thesis that a class has its subclasses as parts, not to mention our usual supposition that non-atomic individuals are sometimes members of classes.

[14] So called because it takes set-theoretical mathematics to consist of generalizations over all the many structurally similar member-singleton relations; not because it posits a new sort of an entity, an 'abstract structure' common to all those relations.

Ramsifying out the Singleton Function

We may replace it by (a translation of) an explicit generalization.

For all S: if ...S...S...S..., then
——S——S——S——

(We might wish to conjoin the Ramsey sentence to the generalization, to prevent vacuous truth in case no suitable relation exists.) If the further sentence followed from the axioms, plus principles and hypotheses of the framework, then also the generalization follows from those auxiliary principles and hypotheses. (Compare first order logic: if $T(s)$ follows from $A(s)$ and $F_1, F_2,...$ and the constant s does not occur in the Fs, then the generalization

For all x: if $A(x)$ then $T(x)$

follows from the Fs alone.)

Structuralism proposes to reconstrue all theorems of set theory – that is, all theorems of the whole of set-theoretical mathematics – as the corresponding generalizations. This reconstrual is a painless reform. No previously accepted theorem must be denied; no previously accepted method of proof must be renounced. Mathematics may go on just as before. Only our understanding of it needs revision.

(And yet structuralism *is* a reform; and it is a reform that serves little purpose if we really did have a primitive understanding of the member-singleton relation; and so it is a rebuke to the old mathematics that was content to take membership as primitive. A modest philosopher might well hesitate to rebuke the old mathematics. He might think that if he cannot understand how it was entitled to its primitive member-singleton relation, most likely the fault is his.)

The philosophical reward of structuralism is that it

Appendix on Pairing

bypasses all doubt about whether the primitive member-singleton relation is well enough understood. You might complain (as Lewis does in section 2.1) that your introductory lesson in set theory just does not apply to the case of membership in a singleton. You might complain that all you were told amounts only to this: it's just like collecting many into one, only without collecting many into one. You might complain that no known theory of intentionality explains how you can have in mind just one out of all the suitable relations. You might complain that there's no hope of reducing the member-singleton relation to any familiar properties and relations, because those familiar properties and relations cannot distinguish inaccessibly many atoms. – Whatever the merit of these complaints, structuralism makes them moot. It says there is no primitive notion that needs understanding; all the suitable relations are equally 'the' member-singleton relation; we needn't try to wrap our minds around some special one of them.

But structuralism is no panacea. It does not put paid to all philosophical complaints that might be brought against set-theoretical mathematics. If you complain not about the 'ideology' but about the ontology of set theory, structuralism doesn't help. Whether the singleton function is primitive or whether it is Ramsified, set theory still requires the hypothesis that there are inaccessibly many atoms. Only very few of them can be ordinary atoms, of the kinds we know and name by causal interaction. The rest are mysterious. We know nothing about their whereabouts, or lack of it. We know nothing about their intrinsic qualitative character, or lack of it. We have to believe in these mysterious extra atoms, on pain of rebelling against established mathematics. Structuralism does nothing to relieve the burden.

In fact, it makes matters worse. We might have been attracted to the speculation that singletons are where their

Ramsifying out the Singleton Function

members are, and perhaps even share the qualitative character of their members. (We might have hoped to answer, somehow, the question how atoms can share the location or character of non-atoms.) But structuralism makes nonsense of this speculation. If all suitable relations alike are member-singleton relations, then a singleton does not have its member once and for all. For anything that can be a member, for anything that can be a singleton, there will be some member-singleton relation that pairs the one with the other. The atom that is Possum's singleton under one suitable relation is Magpie's singleton under another. Shall it share Possum's location and character, or shall it share Magpie's?

If you complain (as Hazen does) about the non-constructiveness of set-theoretical mathematics, again structuralism doesn't help. All it does is redirect your complaint from set theory to plural quantification and mereology. If you will not affirm an existential statement until you see a confirming instance – and all the more if you question whether quantification over unconstructed mathematical objects is meaningful – you should not like the Ramsey sentence of mereologized arithmetic any better than you liked the original axioms. In fact the Ramsey sentence is worse by one initial quantifier – or rather, under translation, by one long string of quantifiers. You should like the Axiom of Choice no better as a principle of the framework than as an axiom of set theory. And you have a second reason to challenge the positing of inaccessibly many atoms, quite apart from any qualms about their unknown location and character.

If you simply complain that the familiar axioms of set theory are not a self-evident foundation for mathematics, again structuralism doesn't help. The Ramsey sentence of mereologized arithmetic is far from self-evident, and the hypothesis that there are inaccessibly many atoms is downright surprising. In the case of the Ramsey sentence, there is some room

Appendix on Pairing

for improvement; in the next section, we shall see how to regain it from a seemingly weakened version. Even so, there is little hope for a self-evident foundation. For better or worse, it is the edifice that justifies the foundation – or nothing does.

Equivalence under Ramsification

Structuralism also brings a mathematical reward. (So it does not quite leave mathematics unchanged. Nothing is lost, but something is gained.) It lets us weaken the axioms of mereologized arithmetic. Without Ramsification, the new, weakened axioms are not equivalent to the old axioms. If a relation satisfies the new axioms, it does not follow that the same relation satisfies the old ones. Yet if some relation satisfies the new axioms, it may follow that *some* relation satisfies the old ones. (The converse is trivial, given that the old axioms imply the new ones.) If so, the old and new Ramsey sentences are equivalent. And if all that we really assert is the Ramsey sentence, as structuralism claims, then the new axioms are interchangeable with the old. There is no substantive question which are right.[15]

Two simultaneous weakenings (at least) are possible. One is to drop the axiom of Induction, and put in its place the much weaker

Atomicity: Every singleton is an atom.

(In section 4.2, Atomicity was derived using Induction.) The other is to retain the positive part of the axiom of Domain

---

[15] For a general discussion of equivalence under Ramsification, see Jane English, 'Underdetermination: Craig and Ramsey', *Journal of Philosophy*, 70 (1973), pp. 453–62.

Equivalence under Ramsification

Domain$^+$: Any part of the null set has a singleton; any singleton has a singleton; any small fusion of singletons has a singleton,

and drop the negative final clause, which says that nothing else has a singleton. Dropping Induction means leaving it open that there might be non-well-founded sets, involved in infinite descents of membership.[16] Dropping the negative clause in the axiom of Domain means leaving it open that there might be scattered exceptions to the principle of Limitation of Size: occasional cases in which a large fusion of singletons has a singleton. (Scattered exceptions only: almost all classes must lack singletons, else there would not be enough singletons to go around; almost all classes are large; therefore almost all large classes must lack singletons.) Dropping the negative clause also means leaving it open that there might be singletons of mixed fusions of individuals and classes, *pace* the invocation of 'our offhand reluctance to believe in them' in section 1.3. If we think we have a primitive understanding of the member-singleton relation, it is a substantive question whether the relation we have in mind really does allow any of these things to happen; whether it satisfies the original axioms of mereologized arithmetic as well as the weakened axioms. But if we are structuralists, and think we did not have any one relation in mind, that question makes no sense. We can weaken the axioms without loss.

Proof Let singleton$_1$ be a relation that satisfies Functionality, Domain$^+$, Distinctness, and Atomicity, but perhaps not the negative clause of Domain and perhaps not Induction. Then there is a relation singleton$_3$ that satisfies all the axioms

---

[16] See Peter Aczel, *Non Well-founded Sets* (Center for the Study of Language and Information, Stanford University, 1988).

Appendix on Pairing

of mereologized arithmetic. (We subscript defined terms of mereologized arithmetic to indicate which singleton relation they correspond to.)

Call a singleton$_1$ *required* iff, whenever there are some atoms such that every singleton$_1$ of an individual$_1$ is one of them, and every singleton$_1$ of one of them is one of them, and every singleton$_1$ of a small fusion of them is one of them, then it is one of them. Atomicity ensures that singletons$_1$ are indeed atoms. Take the class of non-self-members$_1$ whose singletons$_1$ are required. It is large; else it would have a singleton$_1$ by Domain$^+$, and its singleton$_1$ would be required, so it would be a self-member$_1$ iff it were not.

So all the required singletons$_1$ are many, so they are equinumerous with all the singletons$_1$, so there is a relation that maps the required singletons$_1$ one-one onto all the singletons$_1$. Let us extend it to a relation M that also maps all individuals$_1$ onto themselves. Let *singleton$_2$* be the restriction of singleton$_1$ to the required singletons$_1$ and the individuals$_1$, and let *singleton$_3$* be the image of singleton$_2$ under M. It is easily shown that singleton$_3$ satisfies Functionality, Domain, Distinctness, and Induction.   QED

A second use of equivalence under Ramsification is of metaphysical rather than mathematical interest. Recall the discussion of 'unofficial axioms' of set theory in sections 2.1 and 2.6. You might think that the 'structural' axioms of mereologized arithmetic are not really enough to characterize a suitable member-singleton relation. You might want a metaphysical characterization as well. There may be little positive to say, but at least there is a *via negativa*: you can insist that no singleton is part of any material thing, or part of space-time, or part of any platonic form, or part of any spirit (either temporal or eternal), or.... If you think this *via negativa* is an essential ingredient in our conception of singletons, you

Equivalence under Ramsification

could insist on incorporating it into the axioms. The resulting conjunction of axioms will be a strengthening of the axioms of mereologized arithmetic; and the resulting Ramsey sentence will go beyond the vocabulary of the framework. Although the primitive notion 'singleton' will be Ramsified out, such terms as 'material', 'space-time', 'platonic form', 'spirit', . . . will remain. This sort of Ramsey sentence says that there is a relation that is suitable not only mathematically but also metaphysically.[17]

But if structuralism is right, the strengthening may be illusory. Suppose for definiteness that we contemplate adding one 'unofficial' axiom of the form

> *X-Axiom*: An atom is an individual atom iff it is one of the atoms X [somehow specified], otherwise it is a singleton.

And suppose we are prepared to affirm, as an auxiliary hypothesis, that only few atoms are among X, and hence that many atoms are not. (That means it would be foolhardy to specify X as 'the atoms located in space-time' or 'the atoms having some qualitative character'; such a specification would go beyond classifying the atoms, and would advance the daring speculation that some – indeed, many – atoms lack location or character.) Assume, as usual, that Reality is of inaccessible size. Then the X-Axiom is redundant under Ramsification. It is not a substantive issue whether to take it or leave it. We get equivalent Ramsey sentences either way.

---

[17] Further, if we strengthen the axioms by giving a necessary and sufficient condition for an atom to be an individual, then we can apply the result of section 4.8: two member-singleton relations that both satisfy the strengthened axioms differ only by a permutation of singletons.

Appendix on Pairing

Proof Let singleton$_3$ be a relation that satisfies all the axioms of mereologized arithmetic, but perhaps not the X-Axiom. Then there is a relation singleton$_8$ that satisfies the axioms of mereologized arithmetic and the X-axiom as well.[18]

Let n be the fusion of the atoms X and whatever atomless gunk there may be. The X-axiom tells us that n ought to end up as the null set. We have assumed that n is small.

Since we have all of mereologized arithmetic and the hypothesis that Reality is of inaccessible size, we may use set theory freely. There is a rank$_3$, call it R, big enough so that (1) for some atoms Y at rank$_3$ R, Y are equinumerous with the parts of n; and (2) for some atoms Z below rank R, Z are equinumerous with X. Let m be the fusion of the atoms Z and whatever gunk there may be. There is a relation *singleton$_4$* that maps the parts of m one-one onto Y. Say that an atom is *above* Y iff, whenever there are some atoms and every atom of Y is one of them, and every singleton of one of them is one of them, and every singleton of a small fusion of them is one of them, then it is one of them. Let the relation *singleton$_5$* be the restriction of singleton$_3$ to atoms above Y. Let the relation *singleton$_6$* be the union of singleton$_4$ and singleton$_5$. The atoms above Y are many; the atoms above Y together with the atoms Z are likewise many. So we have a relation that maps the atoms above Y together with the atoms Z one-one onto all atoms; and let us extend it to a relation M that also maps all atomless things onto themselves. Let *singleton$_7$* be the image of singleton$_6$ under this M. It is easily shown that singleton$_7$ satisfies all the axioms of mereologized arithmetic. It does not satisfy the X-Axiom (unless by luck);

---

[18] As the subscripts suggest, this proof may be joined to the previous one to show that the weakening of mereologized arithmetic just considered is equivalent under Ramsification (given the specified hypotheses) to the strengthening of mereologized arithmetic which adds an X-Axiom.

Equivalence under Ramsification

however the individual$_7$ atoms are equinumerous with the atoms X that ought to be individual atoms.

So there is a permutation of atoms that maps the individual$_7$ atoms onto X. Let us extend it to a relation N that also maps all atomless things onto themselves. Let *singleton*$_8$ be the image of singleton$_7$ under N. It is easily shown that singleton$_8$ satisfies all the axioms of mereologized arithmetic, and the X-Axiom as well.   QED

Index

acts of set-forming, 29–31
Aczel, Peter, 145
Angelelli, Ignacio, 5
arithmetic, 48, 53–4, 107–13, 113–15
Armstrong, D. M., ix, 5, 56–7, 63, 76, 82–3, 86
Atomicity, Axiom of, 144
atoms, definition of, 15
Aussonderung, Axiom of, 101–2
awesome 'classes', 66–8
axioms, for arithmetic, vii, 48, 107; for mereologized arithmetic, vii, 46, 62, 95–6, 112, 139; for mereology, 74; for ordinal arithmetic, 113; for set theory, 100–7; unofficial, 31–4, 46, 49, 109, 115–16, 146–7

Baxter, Donald, ix, 82–4
Benacerraf, Paul, ix
Bigelow, John, 55

Black, Max, 63
Boolos, George, ix, 19, 52, 62–3, 66, 70
Bricker, Phillip, ix
Bunt, Harry C., vii–viii, 10, 16, 21, 61, 73
Burgess, John P., viii, ix, 53, 104–5, 121–7

Campbell, Keith, 33–4, 76
Cantor, Georg, 5, 27, 29, 105
cardinals, *see* large cardinals, size
Carnap, Rudolf, 47
categoricity, 113–20, 147
character of singletons, 33–5, 57, 142–3, 147
Choice, Axiom of, 71–2, 102, 103–4, 130, 143
class, definitions of, 4, 16, 97
class-like things, 51, 65, 116
coding of pseudo-members, 23–8
completeness, 113–20, 147

Index

composition, as identity, 81–7; uniqueness of, 38–9, 74, 78–9, 85, 100; unmereological, 3, 38–41, 57, 79; unrestricted, 7–8, 18–19, 74, 79–81, 85, 101, 112
connected parts, 22–7
conservatism, ix, 9, 19, 50, 54, 58–9, 141
constructiveness, 143
Continuum Hypothesis, 139
counterparts, 37–8, 77
Cresswell, M. J., ix

Davoren, Jennifer, ix
decomposition, uniqueness or multiplicity of, 2–3, 5, 22, 24
Dedekind, Richard, 58, 88–9; *see also* Peano axioms
difference, mereological, 9, 17
Distinctness, Axiom of, 95, 107–9, 112
Division Thesis, 7–9, 15, 16, 43–4, 99
Domain, Axiom of, 95, 107–9, 112, 144–6
double images, 121–7, 133

Eberle, Rolf, 73
empty set, *see* null set
English, Jane, 144
equivalence under Ramsification, 144–9
Etchemendy, Nancy, ix
Extensionality, Axiom of, 100
external relations, 34–5, 37–8
extraneous orderings, 127–33

facts, 8, 56–7
few and many, definitions of, 90–1, 136
Field, Hartry, ix, 58
finite size, 88–9, 94–5, 96–7, 107
First Thesis, 4–6, 9–10, 56, 98
Fitzgerald, Paul, 45, 139, 140
Forrest, Peter, 56
framework, *see* mereology, plural quantification
Functionality, Axiom of, 95, 107, 112
functions, *see* relations
Fundierung, Axiom of, 18, 20, 103; *see also* Induction
fusion, definitions of, 1–2, 73
Fusion Thesis, 7, 9, 15, 99

Generalized Continuum Hypothesis, 139
generating relations, 38–41
geometry, 113–14
God, 8, 9, 31, 75, 76
Gödel's incompleteness theorem, 114–15
Goldbach's conjecture, 114
Goodman, Nelson, 33, 34, 38–41, 73, 76
gunk, 20–1, 39, 70, 74, 88, 89, 121, 133–6

haecceities, 55, 57
Halmos, Paul R., 29–30
Hazen, A. P., viii, ix, 53, 127–33, 143
hierarchy of classes, 6, 12–13, 20, 26, 27–8, 32, 35
Holmes, Sherlock, 110

Index

Hume, David, 70
Hypotheses I, P, and U, 93–4, 104–7, 134, 148–9

identity, composition as, 81–7
images, 121–7, 133
improper parts, definition of, 1
inaccessible size, 94, 136–7, 142, 143, 147–8
inclusion, 10, 11, 17
individuals, definitions of, 4, 15, 17, 97; as parts of classes, 6, 42–5
Induction, Axiom of, 96, 107, 109, 112, 113, 144–6
infinitary sentences, 50, 69, 70–1
infinite and finite, 88–9, 94–5, 96–7, 106–7
Infinity, Axiom of, 106–7
internal relations, 34–7
Isaacson, Daniel, ix, 30

Jekyll and Hyde, 6, 31, 61
Johnston, Mark, ix

Kleene, Stephen C., 29–30
Kuratowski, Kazimierz, 102, 117, 126

large cardinals, 49, 137–8
large and small, definitions of, 89–91, 136
Lasso Hypothesis, 42–5
Leonard, Henry, 73
Leśniewski, Stanisław, 72–3
Lévy, Azriel, 104–5
Lewis, David, 3, 4, 21, 77, 79, 81, 111
limitation of size, 20, 28, 98, 145

Magpie, 2, 6, 8, 14, 16, 32, 42, 56, 84, 85, 110, 122, 143
Main Thesis, 7, 15–17, 43, 55, 100
Malezieu, Nicolas de, 70
many and few, definitions of, 90–1, 136
Martin, Richard M., 10
Massey, Gerald, 63
mathematics as set theory, ix, 6, 12, 26–7, 36, 53–4, 58, 87, 93, 94, 115, 141, 143–4
maximal connected parts, 22–7
measurable cardinals, 137–8
megethology, 27, 49–50, 74, 93–5, 120, 136–9
membership, definition of, 16, 97
mental acts of set-forming, 29–31
mereologized arithmetic, 95–8
mereology, 1–3, 72–87, and *passim*
metaphysics, 54–7; *see also* unofficial axioms of set theory
microcosms, 121–7
mixed fusions, 7–8, 15, 17, 21, 45, 46, 80, 112, 145
Morton, Adam, 63, 70
mystery about sets, vii, 5–6, 29–38, 41–2, 45, 48–50, 53, 54–7

nice parts, 22–3, 27
nominalism, 21, 65, 66–7
nominalistic set theory, 21–8
non-self-members, 8, 18–19, 27, 63–4, 65–8
null individual, viii, 10–12, 12–13
null set, 4, 10–15, 17, 18, 21, 32–3, 95–6, 97, 98, 100, 139
Null Set, Axiom of, 100

Index

ontological innocence, 62, 68–9, 81–7, 93, 102
Ontology, Leśniewski's, 72–3
Oppy, Graham, ix
orderings, extraneous, 127–33
ordinals, 112–13
overlap, definition of, 73

Pairing, Principle of, 132
pairs, 52–3, 55, 57, 71, 102, 117, 122, 125–7, 131–3, 135
Pair Sets, Axiom of, 100–1
parallel postulate, 113–14, 115
Peano axioms, vii, 48, 95–6, 107–9, 112–13, 114–15, 144–6
permutations of singletons, 116–20
philosophers, follies of, ix, 59
plural quantification, 19, 52, 62–71, 72–3, 87, 101–2, 115, 116, 120, and *passim*
Pollard, Stephen, 63
possibilities, as abstract representations, 35–8, 77–8; as worlds, 13, 36, 37, 49, 77, 86
Possum, 2, 6, 8, 12, 13, 14, 15, 16, 32, 35, 37, 39, 42, 47, 55–6, 82, 84, 85, 98, 109, 110, 122, 143
Power Sets, Axiom of, 93, 104, 105
Prior, A. N., 73
Priority Thesis, 7, 9, 15, 16, 99
proper classes, 4, 9, 15, 18–19, 20, 65, 66–8, 98, 112, 120
proper parts, definition of, 2
properties, 8, 21, 22, 33–4, 55–7, 76, 79, 86; *see also* relations
pseudo-members and pseudo-classes, 21–8

pure classes, 12–13, 32–3, 112, 117
Putnam, Hilary, ix

qualitative character of singletons, 33–5, 57, 142–3, 147
Quine, Willard V., ix, 41–2, 111, 132

Ramsey, Frank P., 47
Ramsey sentences, 46–54, 139–49
rank, 35, 67, 148
Reality, map of, 19–21
relations, existence of, 50–1; generating, 38–41; internal and external, 36–8; quantifying over, viii, 45–54, 121–49; theories of, 51–2
Replacement, 90, 91, 102
Resnik, Michael, 66
Robbin, Joel W., 29–30, 114
Robinson, Denis, ix
Russell, Bertrand, 63; *see also* non-self-members

second-order systems, 63, 66, 70–1, 114, 116, 117
Second Thesis, 6–10, 56
sets, definitions of, 4, 18, 98
Sharvy, Richard, 63
Shoenfield, Joseph R., 29–30, 138
Simons, Peter, 63, 73
singleton, 12, 15, and *passim*; identified with its member, 41–2; as primitive, vii–ix, 16, 31, 45–7, 52, 54, 61, 95, 109–12, 115, 139–42, 145, 147
singularism, 65–9, 70, 116
size, finite and infinite, 88–9,

Index

94–5, 96–7, 106–7; inaccessible, 94, 136–7, 142, 143, 147–8; large and small, 89–92, 136; limitation of, 20, 28, 98, 145; of reality, 27, 49–50, 74, 93–5, 120, 136–9

Skolem, Thoralf, 46, 50, 101

small and large, definitions of, 89–91, 136

space-time, 14, 31–5, 46, 75–6, 78, 85–7, 89, 142–3, 147

states of affairs, 8, 56–7

Stenius, Eric, 63

structuralism, viii–ix, 45–54, 58, 110–11, 139–49

substitutional 'quantification', 50, 51–2, 64–5

successor relation, 48, 49, 53–4, 107–13, 114

supervaluations, 14, 47

tandem, mappings in, 123–5

Tarski, Alfred, 73

Taylor, Barry, ix

time, *see* space-time

transitivity, 1, 3, 5, 74

Trisection, Principle of, 124

tropes, 8, 33–4, 51, 55–6, 57, 76

unicle relation, 61; *see also* singleton

union, 18

Unions, Axiom of, 93, 104–5

Uniqueness of Composition, 38–9, 74, 78–9, 85, 100

unit class, *see* singleton

unithood facts, 56–7

universals, 8, 21, 22, 33–4, 51, 76, 79, 86

unofficial axioms of set theory, 31–4, 46, 49, 109, 115–16

Unrestricted Composition, 7–8, 18–19, 74, 79–81, 85, 101, 112

urelements, 4, 15, 41–2, 98

van Fraassen, Bas, 14, 47

van Inwagen, Peter, ix, 35–8, 81

von Neumann, John, 98, 104, 111, 113

Well-ordering Principle, 130

Weston, Thomas, 116, 117

Wiener, Norbert, 126

Williams, Donald C., 33, 56, 76

Zermelo, Ernst, 5, 110–11